NIST Special Publication 250-69

NIST MEASUREMENT SERVICES:
Regular Spectral Transmittance

David W. Allen, Edward A. Early, Benjamin K. Tsai, and Catherine C. Cooksey

Optical Technology Division
Physical Measurement Laboratory
National Institute of Standards and Technology
Gaithersburg, MD 20899-0001

March 2011

U.S. DEPARTMENT OF COMMERCE
Gary F. Locke, Secretary

National Institute of Standards and Technology
Patrick D. Gallagher, Director

PREFACE

The calibration and related measurement services of the National Institute of Standards and Technology are intended to assist the makers and users of precision measuring instruments in achieving the highest possible levels of accuracy, quality, and productivity. NIST offers over 300 different calibrations, special tests, and measurement assurance services. These services allow customers to directly link their measurement systems to measurement systems and standards maintained by NIST. These services are offered to the public and private organizations alike. They are described in NIST Special Publication (SP) 250, NIST Calibration Services Users Guide.

The Users Guide is supplemented by a number of Special Publications (designated as the "SP250 Series") that provide detailed descriptions of the important features of specific NIST calibration services. These documents provide a description of the: (1) specifications for the services; (2) design philosophy and theory; (3) NIST measurement system; (4) NIST operational procedures; (5) assessment of the measurement uncertainty including random and systematic errors and an error budget; and (6) internal quality control procedures used by NIST. These documents will present more detail than can be given in NIST calibration reports, or than is generally allowed in articles in scientific journals. In the past, NIST has published such information in a variety of ways. This series will make this type of information more readily available to the user.

This document, SP250-69 (2011), NIST Measurement Services: Regular Spectral Transmittance is a new publication. It covers the special tests of spectral transmittance (test number 38061S in SP250, NIST Calibration Services Users Guide), the calibration of spectral transmittance filters (test numbers 38010C through 38040C), and survey scans for research and development. Inquiries concerning the technical content of this document or the specifications for these services should be directed to the author or to one of the technical contacts cited in SP250.

NIST welcomes suggestions on how publications such as this might be made more useful. Suggestions are also welcome concerning the need for new calibrations services, special tests, and measurement assurance programs.

James K. Olthoff
Deputy Director for Measurement Services
Physical Measurement Laboratory

Katharine B. Gebbie
Director
Physical Measurement Laboratory

ABSTRACT

This document describes measurement services, instrumentation, and techniques for regular spectral transmittance over the spectral range from 250 nm to 2500 nm at the National Institute of Standards and Technology. Section 2 explains the basic theory in deriving the transmittance, while a more thorough derivation is reserved for the interested reader in Appendix A. Section 3 describes the reference instrument hardware, software, validation, procedures, and uncertainties. Section 4 provides this same detailed information for the transfer instrument. Sample calibration reports and sample data are provided in Appendices B and C, respectively, and Appendix D details instructions for requesting Regular Spectral Transmittance Calibrations. The measurement services are described for the benefit of customers and potential customers, and the instruments, measurement procedures, and uncertainties are described in sufficient detail to allow comparisons with measurement methods at other laboratories.

Key words: calibration, regular spectral transmittance, transmittance, uncertainty

TABLE OF CONTENTS

Abstract 2

1. Introduction 6

2. Transmittance Measurement Equation 7

3. Reference Instrument 8
   3.1 Hardware 8
      3.1.1 Illuminator 9
      3.1.2 Sample Translation 10
      3.1.3 Receiver 10
      3.1.4 Geometrical Optics 11
      3.1.5 Spectral Characterization 11
      3.1.6 Electrical 19
   3.2 Instrument Validation 20
   3.3 Measurement Procedure 20
   3.4 Data Acquisition 21
   3.5 Data Analysis 22
   3.6 Uncertainties 23
   3.7 Calibration Report 25

4. Transfer Instrument 26
   4.1 Instrument Description 26
   4.2 Instrument Calibration 27
   4.3 Measurement Procedure 27
   4.4 Data Analysis 28
   4.5 Uncertainties 28

Acknowledgments 28

References 29

Appendix A: Transmittance Theory and Measurement Equation 30
Appendix B: Sample Report 43
Appendix C: Sample Data 55
Appendix D: How to Request Calibration Services 57

LIST OF FIGURES

Figure 1. Schematic of the RTS 9

Figure 2. Slit-scattering function measured with emission line lamps for the indicated grating and slit width 12

Figure 3. Normalized signal from Ar^+ laser used to measure the stray-light rejection of the monochromator 13

Figure 4. Signal as a function of wavelength with the D_2 arc lamp source and the PMT photomultiplier detector (a) with and (b) without the polarizer in the influx beam and 2 mm wide slits on the monochromator 15

Figure 5. Signal as a function of wavelength with the QTH incandescent lamp source and the Si photodiode without integrating sphere and 1 mm wide slits on the monochromator 16

Figure 6. Signal as a function of wavelength with the QTH incandescent lamp source and the Si photodiode without integrating sphere and 2 mm wide slits on the monochromator 17

Figure 7. Signal as a function of wavelength with the QTH incandescent lamp source and the Si photodiode with integrating sphere and 2 mm wide slits on the monochromator 18

Figure 8. Signal as a function of wavelength with the QTH incandescent lamp source and the extended-range InGaAs photodiode with integrating sphere and 2 mm wide slits on the monochromator 19

Figure A1. Interactions of incident radiation with a surface boundary 31

Figure A2. Interactions of incident radiation within a layer 35

Figure A3. Transmittance as a function of incident angle for the indicated polarizations of the incident radiant flux of a sample with an index of refraction of 1.5, thickness of 2 mm, and attenuation coefficient of 0.005 mm^{-1} 36

Figure B1. Regular spectral transmittance τ as a function of wavelength λ of neutral density filters 46

Figure C1. Transmittance as a function of wavelength of the color filters 55

Figure C2. Transmittance, on a logarithmic scale, as a function of wavelength of the MAP (Measurement Assurance Program) filters 55

Figure C4. Measured spectral transmittance of the holmium oxide glass filter for a spectral bandwidth of 1 nm 56

LIST OF TABLES

Table 1. Optical detector choices based on wavelength 11

Table 2. Geometrical specifications of the RTS 11

Table 3. Spectral bandwidth used to measure the slit-scattering function 13

Table 4. Spectral specifications of the RTS 14

Table 5. Gain ratios, for the indicated photodiode and gains, with uncertainties 20

Table 6. Sources of uncertainty and classifications for transmittance measurements 23

Table 7. Sources of uncertainty and typical uncertainty contributions for transmittance 25

Table 8. Sources of uncertainties for various detector systems 25

Table 9. Certified and measured wavelengths of minimum transmittance of a holmium oxide glass filter, for a spectral bandwidth of 1 nm 27

Table B1. Regular spectral transmittance τ as a function of wavelength λ of a neutral density filter, serial number F1-001 47

Table B2. Regular spectral transmittance τ as a function of wavelength λ of a neutral density filter, serial number F1-002 48

Table B3. Regular spectral transmittance τ as a function of wavelength λ of a neutral density filter, serial number F1-003 49

Table B4. Regular spectral transmittance τ as a function of wavelength λ of a neutral density filter, serial number F2-001 50

Table B5. Regular spectral transmittance τ as a function of wavelength λ of a neutral density filter, serial number F2-002 51

Table B6. Regular spectral transmittance τ as a function of wavelength λ of a neutral density filter, serial number F2-003 52

Table B7. Uncertainty contributions and expanded uncertainty ($k = 2$) of the regular transmittance of a neutral density glass filter, serial number F1-001 53

Table B8. Uncertainty contributions and expanded uncertainty ($k = 2$) of the regular transmittance of a neutral density glass filter, serial number F1-002 53

Table B9. Uncertainty contributions and expanded uncertainty ($k = 2$) of the regular transmittance of a neutral density glass filter, serial number F1-003 53

Table B10. Uncertainty contributions and expanded uncertainty ($k = 2$) of the regular transmittance of a neutral density glass filter, serial number F2-001 54

Table B11. Uncertainty contributions and expanded uncertainty ($k = 2$) of the regular transmittance of a neutral density glass filter, serial number F2-002 54

Table B12. Uncertainty contributions and expanded uncertainty ($k = 2$) of the regular transmittance of a neutral density glass filter, serial number F2-003 54

Table C3. Nominal transmittance values at 548.5 nm and materials for the MAP filters 56

1. Introduction

NIST maintains and operates the reference instrument that establishes the national scale for regular spectral transmittance. It is within the NIST mission to maintain this national scale and disseminate it for the benefit of industry, government agencies, and others who need the highest accuracy measurements. Calibration services performed by the NIST Regular Transmittance Laboratory are intended to provide traceability to the national scale and are not intended to be used for meeting generic measurement needs.

The calibration services for regular spectral transmittance are an integral part of the NIST spectrophotometry program, which includes calibration services for regular spectral transmittance, diffuse transmittance, bidirectional reflectance, and directional-hemispherical reflectance. For further information on these and other calibration services, see the existing SP250 documents [1-3] listed inside the back cover of this document.

This laboratory only provides calibrations of regular spectral transmittance for solid samples, such as optical filters. It does not provide services for liquids in glass or fused silica cells. Optical filters can be classified as either neutral density filters or spectrally selective filters. The neutral density filters transmit light approximately equally across a given broad spectral range, while spectrally selective filters transmit light with marked spectral dependence. Such filters include colored glass, band-pass, and interference filters.

The calibration services provided by this laboratory are not intended to be the only means of traceability to the national scale for spectral transmittance. Customers are encouraged to explore the use of secondary laboratories that provide similar services. Often those laboratories may be a more economical solution and provide acceptable measurement uncertainties.

NIST does not state an expiration date for the calibration. The frequency of calibration for an item is determined by the customer. It should be based on the quality of the sample and how the sample is handled and maintained. High quality glass filters can be stable for many years. If the sample is cared for properly, a calibration interval of several years or greater may be reasonable.

NIST is regularly updating services in the spectrophotometry program. For the latest information visit the spectrophotometry program website at http://www.nist.gov/pml/div685/grp03/spectrophotometry.html.

Services Provided

Measurement services are available for dissemination of the scales of transmittance maintained by the reference instrument. The primary purpose of the calibration services is to provide traceability to the scale, not for research on particular test items. Survey scans are available at a reduced cost to customers who do not need traceability to NIST. NIST provides the following calibration services:

Special Tests of Spectral Transmittance (38061S)

Measurements of spectral transmittance can be made at wavelengths from 250 nm to 2500 nm for non-fluorescent submitted test items. In general, measurements are performed at room temperature with the incident beam normal to the front surface of the test item. Items with dimensions from 1 cm to 5.1 cm across perpendicular to the incident beam are routinely accommodated. Other sizes and measurement conditions are possible in consultation with NIST staff. In general, for calibration service 38061S the customer supplies the test item, which is measured under the spectral conditions specified by the customer. Uncertainty estimates will be given and will depend on the optical characteristics of the submitted test item and the instrument used to perform the measurement. Arrangements for these measurements on submitted test items must be made before shipment. The decision to perform the measurements and selection of the instruments to be used will rest with NIST. Test items not accepted for measurement will be returned.

Spectral Transmittance Filters (38010C to 38040C)

NIST calibrates standards of spectral transmittance for verifying the photometric scale of spectrophotometers. Limited quantities are available for purchase with the calibration service. These standards are either 30 mm diameter polished glass disks or 51 mm polished glass squares, 2 mm to 3 mm thick, designated as cobalt blue (38010C), copper green (38020C), carbon yellow (38030C), and selenium orange (38040C). The relative expanded uncertainty ranges from approximately 0.2 % to 0.3 % of the transmittance value. Information provided to the user includes values of transmittance at 25 °C at 10 nm intervals from 380 nm to 770 nm, the estimated uncertainty of each value, and wavelength-dependent temperature coefficients to estimate the variation of the transmittance with change in temperature.

Wavelength Standards (38050C-38051C)

NIST no longer provides a calibration service for holmium oxide glass standards. After decades of observations, NIST determined that holmium oxide glass standards are inherently stable. Users of holmium oxide glass wavelength standards, either provided by NIST or elsewhere, can compare the spectra obtained from transmission measurement and verify that the sample is holmium oxide. If the sample matches the spectra shown in Appendix C, Fig. C4, the user can self declare traceability for wavelength accuracy. Further details can be found in Ref. [4].

The holmium oxide filters are suitable only as wavelength standards and not as transmittance standards. Therefore, NIST does not provide transmittance measurements for these filters.

2. Transmittance Measurement Equation

The transmittance τ of a sample illuminated by unpolarized incident radiant flux is given by

$$\tau(\lambda) = \frac{\tau(\lambda, 0°) + \tau(\lambda, 90°)}{2}, \tag{1}$$

where λ is the wavelength, 0° refers to 0° polarization, and 90° refers to 90° polarization. The average of the two orthogonal polarization states is assumed to be equivalent to the transmittance using unpolarized light. Equation (1) states that the measured sample transmittance is determined by calculating the average of the two polarizations of the transmitted radiant flux at 0° and 90°. At each polarization, σ, the transmittance is determined by the ratio,

$$\tau(\lambda,\sigma_i) = \frac{I_t(\lambda,\sigma_i)}{I_i(\lambda,\sigma_i)} = \frac{S_t(\lambda,\sigma_i)}{S_i(\lambda,\sigma_i)} \cdot \frac{G_i(\lambda,\sigma_i)}{G_t(\lambda,\sigma_i)} ,\tag{2}$$

where I is the intensity, S is the measured signal, G is the transimpedance amplifier gain, and the subscripts i and t denote incident and transmitted, respectively. It is implicitly assumed that there is no change in the polarization state of the radiant flux that is transmitted by the sample and that the detector is insensitive to the polarization of the radiant flux. For a more detailed discussion, see Appendix A.

3. Reference Instrument

The Reference Transmittance Spectrophotometer (RTS) is the national reference instrument for regular spectral transmittance measurements of non-fluorescent samples at room temperature (23 °C ± 2 °C). The RTS performs absolute measurements of the regular transmittance on samples with widths from 1 cm to 5.1 cm. The measurements are usually performed with a collimated incident beam at normal incidence, meaning that the illumination axis is parallel to the normal of the sample. The spectral range for the measurements is 250 nm to 2500 nm. For neutral samples, the smallest measurable transmittance is 10^{-6}. For narrow-band samples, such as interference filters, the smallest measurable transmittance is 10^{-4}.

Unless requested by the customer, the measurements will be performed with the sample centered on the influx beam. The diameter of the influx beam is variable from 4 mm to 20 mm, with 20 mm as the typical diameter.

Filter uniformity is not routinely measured. Arrangements can be made to measure the sample at several spatial positions at one wavelength for an additional cost. NIST may also be helpful in suggesting other means for obtaining uniformity information.

3.1 Hardware

The RTS is located in Building 220, Room A324 on the NIST Gaithersburg campus. The RTS is a custom-built instrument which consists of an illuminator, sample mount, and receiver on a vibration-isolation table. The instrument is enclosed by a black light-tight box with side panels and forced filtered-air ventilation. Electronics are located on top of the box, and an adjacent computer provides automated instrument control, data acquisition, and analysis. All critical electronics are maintained on a 14 kVA uninterruptible power system to eliminate data errors and down time from line-voltage fluctuations. A schematic diagram of the RTS is shown in Fig. 1, indicating the major systems and components.

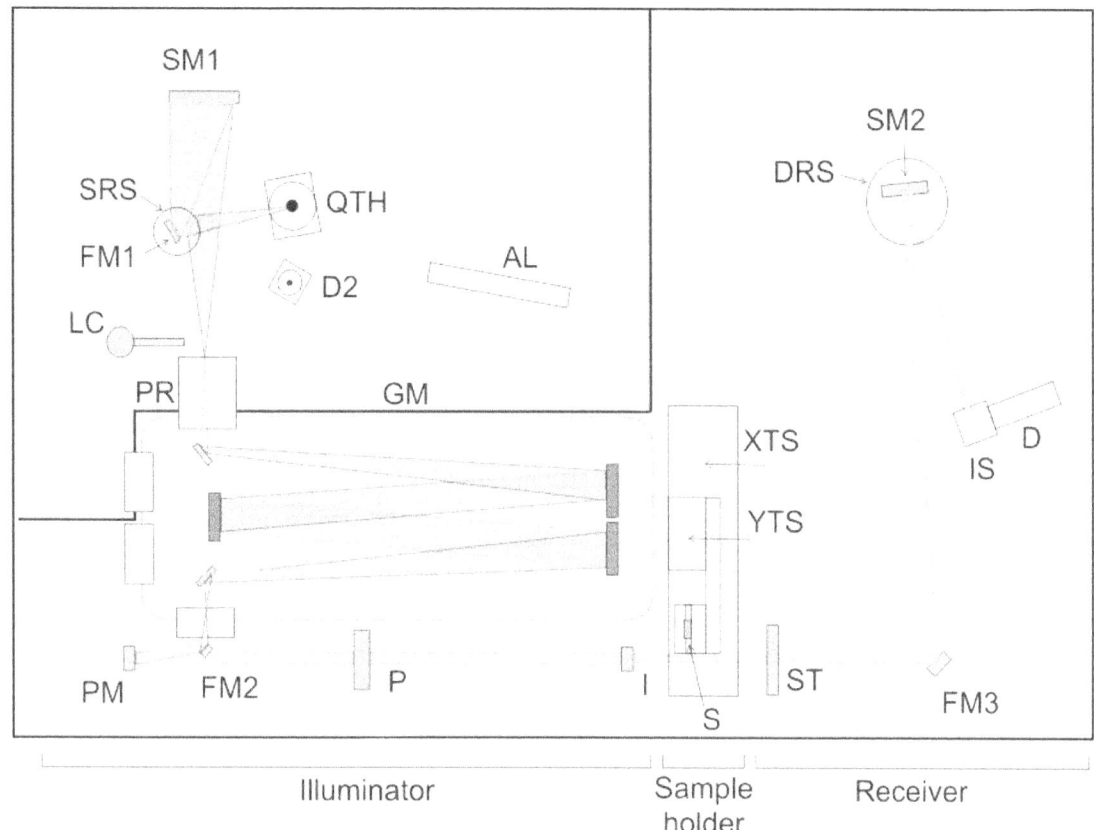

Figure 1. Schematic of the RTS with the systems and components labeled. The light path from source to detector is shaded. The Illuminator system consists of QTH – quartz-tungsten-halogen lamp, D2 – deuterium lamp, AL – alignment laser, FM1 – flat mirror, SRS – source rotation stage, SM1 – spherical mirror, LC – light chopper, PR – prism monochromator, GM – grating monochromator, FM2 – flat mirror, PM – parabolic mirror, P – polarizer, and I – iris. The Sample system consists of S – sample holder, XTS – x-translation stage, and YTS – y-translation stage. The Receiver system consists of ST – shutter, FM3 – flat mirror, DRS – detector rotation stage, SM2 – spherical mirror, IS – integrating sphere, and D – detector.

3.1.1 Illuminator

The illuminator provides a monochromatic, polarized influx beam centered on the sample, with a triangular spectral bandwidth determined by the monochromator slit function. It consists of a radiation source, mirrors, light chopper, prism-grating monochromator, polarizer, and iris. The source and the monochromator's grating and slit settings are selectable.

The two possible sources are a quartz-tungsten-halogen (QTH) incandescent lamp and a deuterium arc lamp. In general, the deuterium lamp is used for wavelengths less than 350 nm and the QTH lamp is used for longer wavelengths. A water circulator cools the housing surrounding the QTH lamp to prevent overheating.

The appropriate source is selected using a rotating flat mirror. The radiant flux is directed towards a spherical mirror with a 15 cm diameter and 28 cm focal length, which focuses the flux from the source onto the entrance slit of the monochromator at a distance of 570 mm from the source. A light chopper can be placed in front of the entrance mask as needed for use with the extended indium-gallium-arsenide (InGaAs) detector.

The f/8.7 monochromator is a 1-m prism-grating system with a two-grating turret and remote selection of the entrance port, entrance and exit slit width, and grating. For wavelengths shorter than 1100 nm, a grating with 600 lines/mm blazed at 200 nm is used. For longer wavelengths, a grating with 600 lines/mm blazed at 1250 nm is used. Individual motors move the prism and gratings, so their wavelengths can be set independently. An encoder is attached to the grating shaft to determine the wavelength, while the wavelength of the prism is determined by the number of motor steps from the home position. The gratings are mounted on a vertical translation stage. The approximate bandwidth, $\Delta\lambda$, determined by the slit width and grating, is given by

$$\Delta\lambda = (1.5 \text{ nm/mm})\left(\frac{600 \text{ lines/mm}}{GD}\right)(SW), \qquad (3)$$

where GD is the groove density in [lines/mm] and SW is the slit width in [mm]. For a 1 mm slit width and a 600 lines/mm grating, the spectral bandwidth is 1.5 nm.

After the exit slit of the monochromator, the optical path expands a distance of 177 mm to a parabolic mirror ($f = 177$ mm), which collimates the beam before arriving at the polarizer. The polarizer is a 2 cm square, uncoated Glan-Taylor prism made of calcite, which has an extinction ratio of greater than 1:100000 and transmits light from 250 nm to 2300 nm. For cases where the polarizer is not suitable (such as when measuring at wavelengths outside practical spectral ranges), the sample is rotated 90° for repeat measurements.

3.1.2 Sample Translation

The sample location is determined by the positions of two translation stages, one for horizontal (x-axis) movement and the other for vertical (y-axis) movement. A platform on the arm attached to the y-axis stage accommodates a variety of sample holders, chosen based upon the size and shape of the sample.

3.1.3 Receiver

The receiver measures both the influx and efflux. The receiver consists of a shutter, mirrors, an integrating sphere, and a detector. The optical axis of the receiver is aligned to the optical axis of the illuminator. The shutter, used to collect the dark current, has a diameter of 65 mm and is coated black to limit scattered light reaching the detector. The integrating sphere, used with the silicon (Si) and the extended-range InGaAs detectors, is lined with polytetrafluoroethylene (PTFE) and has an entrance port of 1 cm and an inside diameter of 5 cm. A baffle is placed in front of the integrating sphere to further reduce the effect of scattered light on the measurements. The entrance port of the integrating sphere is located approximately at the focus of the spherical mirror, so that an image of the exit slit of the monochromator is formed at the port.

Several optical detectors are used, and the choice depends on the wavelength range, as detailed in Table 1. The photomultiplier-tube (PMT) is a module attached to its own integrating sphere, which is also lined with PTFE and has an entrance port of 3 cm and an inside diameter of 10 cm. The high-voltage of the PMT is controlled by a potentiometer, and the output voltage is the current multiplied by 10^6 V/A. The Si detector is temperature-controlled at 26° C with a transimpedance amplifier having gains that are selected manually or remotely. The extended-range InGaAs detector, temperature-controlled at −40° C, contains a transimpedance amplifier with gains that are selected manually or remotely. The output current of the extended-range InGaAs detector passes through a lock-in amplifier.

Table 1. Optical detector choices based on wavelength

Detector	Wavelength Range [nm]
PMT	200 to 400
Si	250 to 1100
Extended-range InGaAs	900 to 2500

3.1.4 Geometrical Optics

The optical system of the monochromator in Fig. 1 is designed to provide adequate and appropriate illumination for the transmittance samples. The degree of collimation determines the illumination or beam quality. Table 2 shows a summary of the geometrical specifications of the RTS, including the degree of collimation or maximum angular deviation of the incoming ray from the receiver axis for both the illuminator and the receiver.

Table 2. Geometrical specifications of the RTS

Property	Illuminator	Receiver
Type of geometry	Directional	Directional
Direction of axis	0° (typical)	180° (typical)
Region	31 mm maximum	35 mm by 50 mm
Sampling Aperture	Influx aperture	
Cone half-angle as measured from PM in Fig. 1	7° (typical maximum)	1.9° by 2.7°
Maximum deviation from collimation at sample	0.16° to 0.81° (depending on slit)	Same as for illuminator

3.1.5 Spectral Characterization

Slit Scattering Function

The slit-scattering functions measured with emission lamps for the various gratings and slit widths are shown in Fig. 2, and the spectral bandwidth (FWHM) for each measurement of the slit-scattering function is given in Table 3. For both slit widths, the slit-scattering function is triangular. For a 1 mm slit width the spectral bandwidth is approximately 1.5 nm, while for the 2 mm slit width it is approximately 3.0 nm. The gratings listed in Fig. 2 and Table 3 are located in the automated turret. Grating A covers the visible range, while Grating B is used for the infrared region.

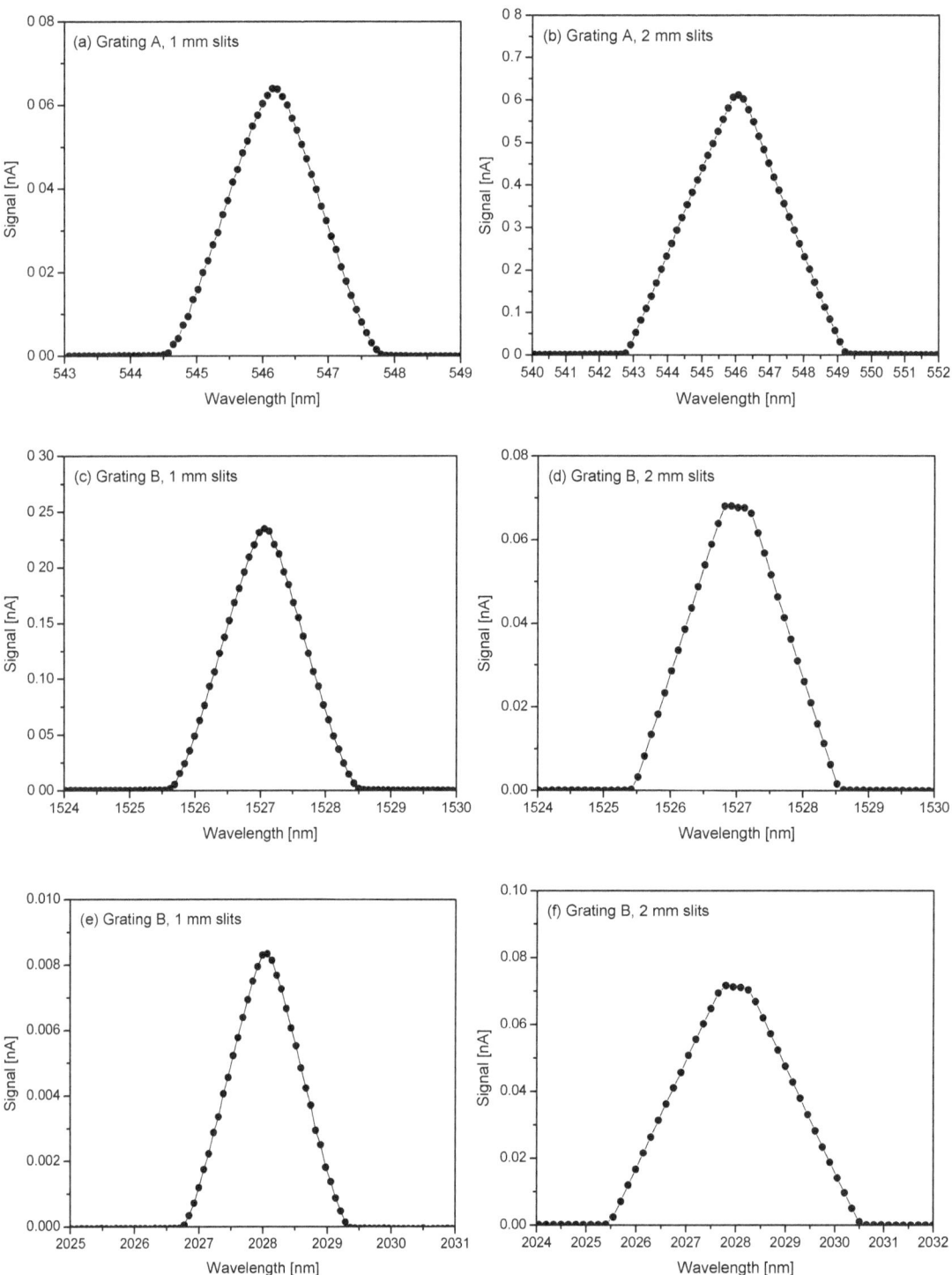

Figure 2. Slit-scattering function measured with emission line lamps for the indicated grating and slit width.

Table 3. Spectral bandwidth for each slit width, emission line, and grating used to measure the slit-scattering function

Slit Width [mm]	Lamp	Line [nm]	Grating	Order	FWHM [nm]
1	Hg	546.1	A	1	1.6
1	Ar	763.5	B	2	1.4
1	Hg	1014	B	2	1.3
2	Hg	546.1	A	1	3.3
2	Ar	763.5	B	2	3.1
2	Hg	1014	B	2	2.8

Stray Light Rejection

The stray-light rejection was measured by operating an Ar^+ laser on six lines simultaneously, as shown in Fig. 3. The stray light was estimated by readings at a wavelength offset from the laser line wavelengths, normalized against the sum of readings at the six laser line wavelengths. An integrating sphere was placed at the entrance of a 2 mm entrance slit. The beam from an Ar^+ laser was projected into the entrance port of the sphere. No polarizer was used in the beam path. The stray-light rejection, approximated as the ratio of the out-of-band signal to the in-band signal, is estimated to be about 10^6.

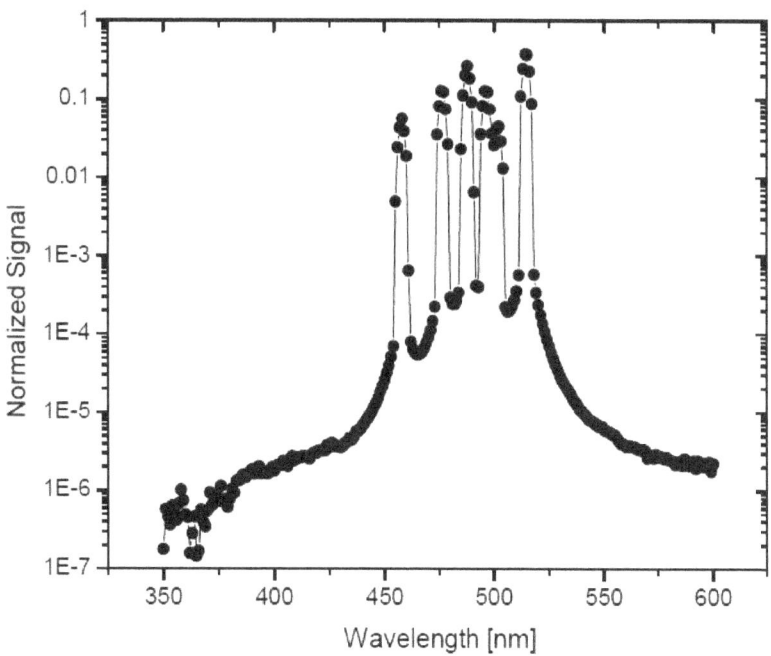

Figure 3. Normalized signal from Ar^+ laser used to measure the stray-light rejection of the monochromator.

Wavelength Calibration

The wavelength was calibrated using Hg, Ne, and Ar emission lamps for both gratings and slit widths of 1 mm and 2 mm. The wavelength uncertainty ($k = 1$) is 0.05 nm for the 1 mm slit width and 0.10 nm for the 2 mm slit width for both gratings. For each line, this uncertainty was estimated from the maximum deviation between known atomic lines to the measured wavelength, or the difference between the calculated and the observed values. A summary of the spectral specifications of the illuminator is given in Table 4.

Table 4. Spectral specifications of the RTS

Property	Value
Wavelength Range	250 nm to 2500 nm
Spectral Bandwidth	1.5 nm, triangular (1 mm slits)
	3.0 nm, triangular (2 mm slits)
Stray-light Rejection	10^6
Wavelength Uncertainty	0.05 nm (1 mm slits)
	0.10 nm (2 mm slits)
Polarization	0° (horizontal) to 90° (vertical)

Signal as a Function of Wavelength

The signal as a function of wavelength for all combinations of sources and detectors is given in Figs. 4 through 8. The spiky structure observed at 0° polarization for several of the signal distributions is due to Wood's anomalies [5].

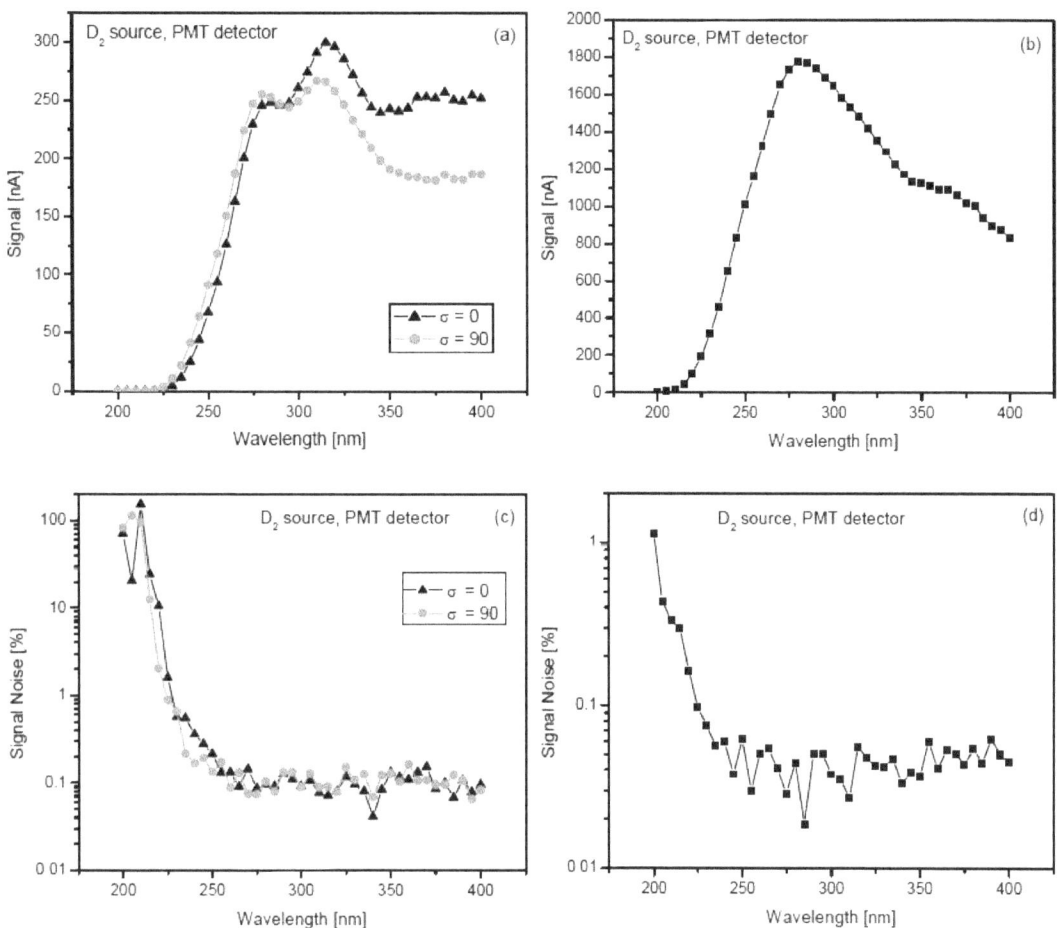

Figure 4. Signal as a function of wavelength with the D_2 arc lamp source and the PMT photomultiplier detector (a) with and (b) without the polarizer in the influx beam and 2 mm wide slits on the monochromator. The signal noise as % of signal for these two conditions is shown in (c) and (d).

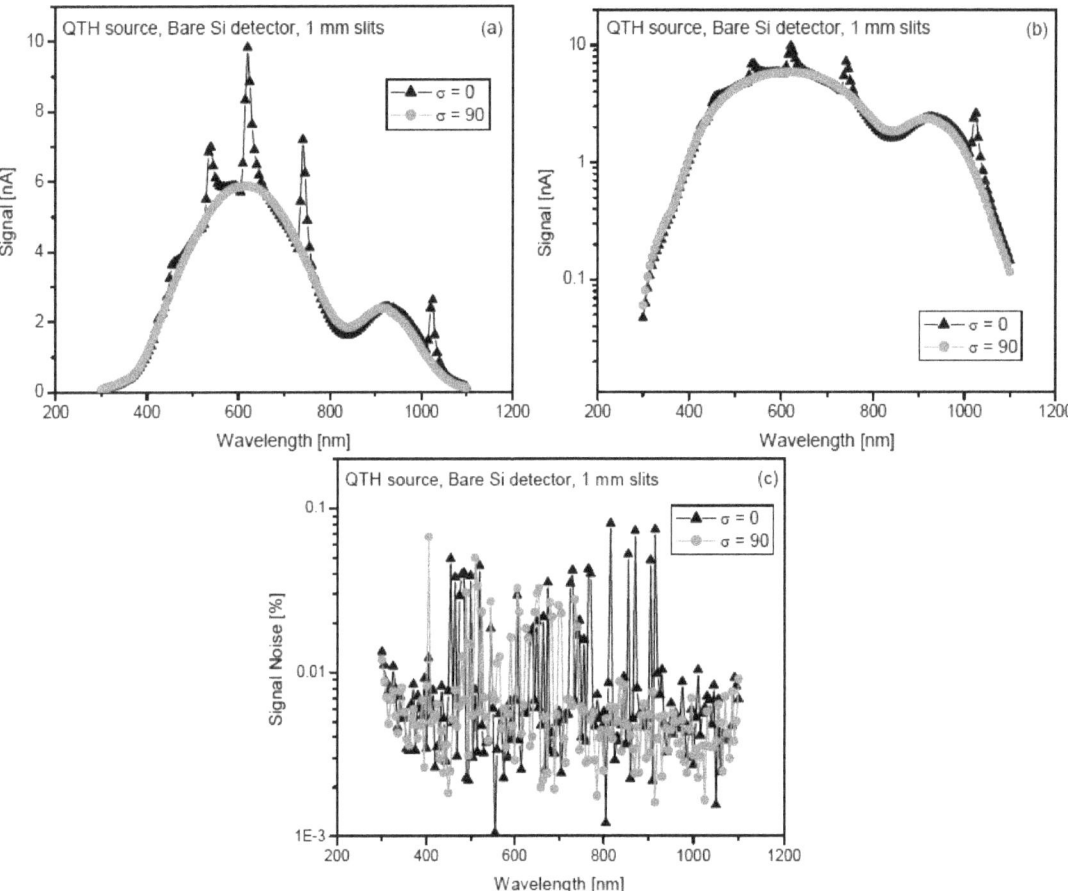

Figure 5. Signal as a function of wavelength with the QTH incandescent lamp source and the Si photodiode without integrating sphere and 1 mm wide slits on the monochromator on an (a) linear and (b) logarithmic scale. The signal noise as % of signal is shown in (c).

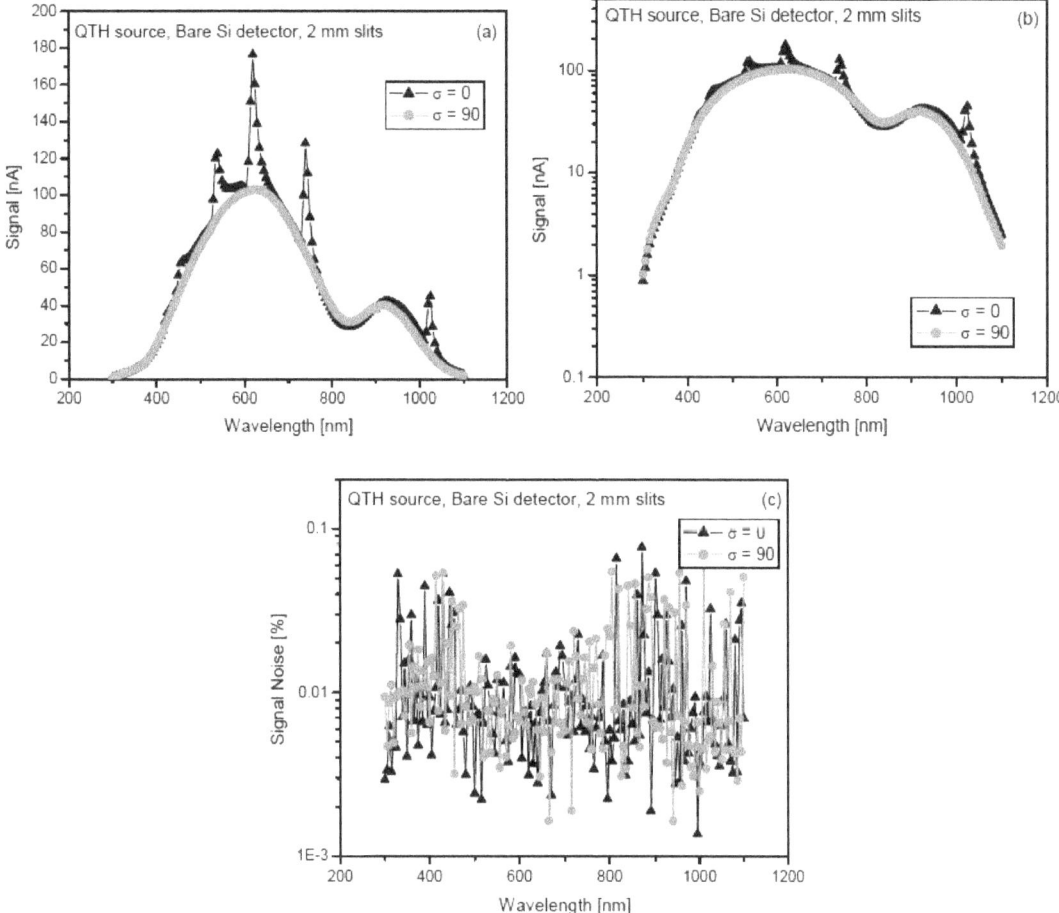

Figure 6. Signal as a function of wavelength with the QTH incandescent lamp source and the Si photodiode without integrating sphere and 2 mm wide slits on the monochromator on an (a) linear and (b) logarithmic scale. The signal noise as % of signal is shown in (c).

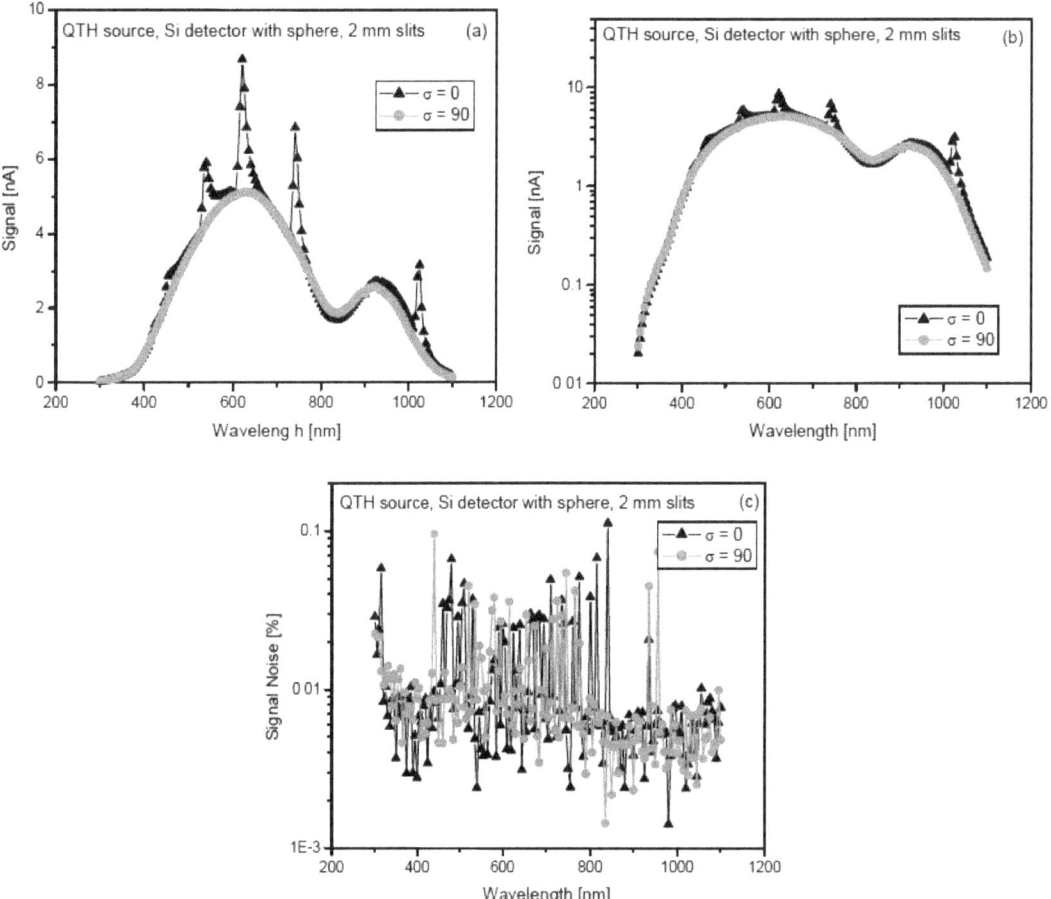

Figure 7. Signal as a function of wavelength with the QTH incandescent lamp source and the Si photodiode with integrating sphere and 2 mm wide slits on the monochromator on an (a) linear and (b) logarithmic scale. The signal noise as % of signal is shown in (c).

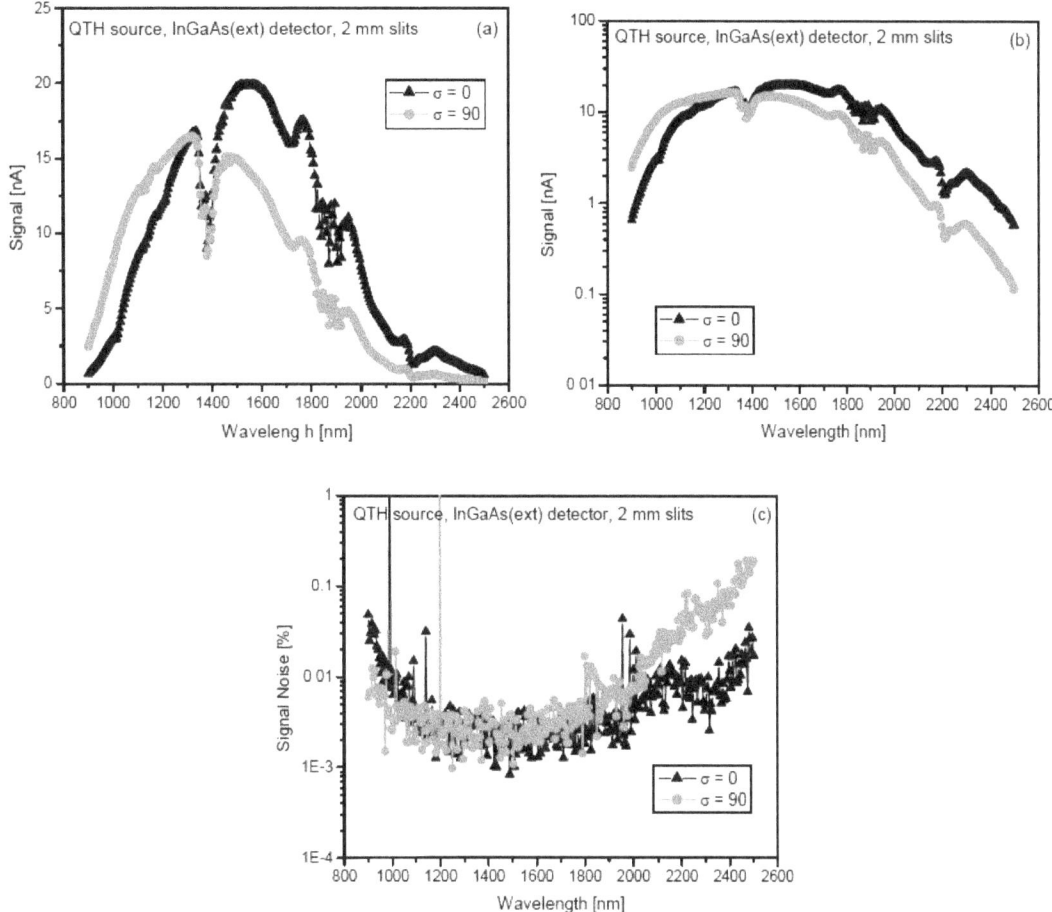

Figure 8. Signal as a function of wavelength with the QTH incandescent lamp source and the extended-range InGaAs photodiode with integrating sphere and 2 mm wide slits on the monochromator on an (a) linear and (b) logarithmic scale. The signal noise is shown in (c).

3.1.6 Electrical

For all detectors, the output current is converted to voltage using a transimpedance amplifier before being measured by the digital voltmeter or lock-in amplifier. The Si and extended InGaAs detectors have a selectable gain on the transimpedance amplifier. Both the voltmeter and lock-in amplifier have ranges, determined by the signal. Since the transmittance involves ratios of signals, it is necessary to determine if the signal is proportional to the current from the detector. This involves measuring the linearity of the signal within a range and gain setting, the range ratio between range settings, and the gain ratio between gain settings, where appropriate. The first assures that the signal is proportional to the current for signals within the range; the second assures that the signal measured in one range is the same as that measured in another range; and the third assures that the signal measured at one gain is the same as that measured at another gain. Ideally, the signal will be linear within a range, the range ratio will be one, and the gain ratio will be 0.1.

The linearity is defined as the ratio of an equivalent signal at two different gain levels.

and is determined by the double-aperture method [6]. All that is required is that the detection system is linear over the full range of light possible through the monochromator with maximum slit width and without the sample. All settings used within this range must be tested. Nonlinearity, *NL*, is defined by

$$NL(\%) = 100 \cdot \left[\frac{\frac{R(I_1)}{G_1} - \frac{R(I_0)}{G_0}}{\frac{R(I_0)}{G_0}} \right], \quad (4)$$

where $R(I_0)$ [V/A] and G_0 are the responsivity and gain at some reference detector current level and $R(I_1)$ is the responsivity [V/A] at some other gain and current level.

The PMT was linear to within 0.2 %; the Si photodiode was linear to within 0.05 %; and the extended range InGaAs photodiode was linear to within 0.075 %. For the digital voltmeter the range ratios were one to within 0.02 %. The range ratios for the lock-in amplifier were one to within 0.02 % with the Si photodiode. The gain ratios and the relative uncertainties for the photodiodes are given in Table 5.

Table 5. Gain ratios for the indicated photodiode and gains, with relative uncertainties

Photodiode	Gains		Gain Ratio	Uncertainty [%] $k = 1$
Si	7	8	0.1	0.01
	8	9	0.1	0.10
Extended InGaAs	7	8	0.1	0.07
	8	9	0.1	0.07

3.2 Instrument Validation

The RTS facility uses three means of measurement assurance: instrument characterization, measurement of check standards, and comparison with other laboratories. Characterization includes assessing the wavelength accuracy and validating the photometric scale.

A full instrument characterization is performed on an annual basis. Check standards are used over the course of a year between measurements to assure the instrument is properly operating. For assessing wavelength accuracy, emission lamps are used. Intercomparisons and reproducibility studies employing neutral-density and colored filters are used for validating the photometric scale.

Control charts track changes in instrument performance over days and years. Data plotted in the control chart include transmittance of check standards at specific wavelengths.

3.3 Measurement Procedure

A measurement is performed by aligning the sample, configuring the instrument for the desired measurement, entering the parameters in the data acquisition program, and running the program to acquire and analyze the data. For each polarization and wavelength, a signal is measured, which is proportional to the efflux. To cancel the effect of source drift to first order, an average of the clear readings bracketing the transmitted reading is used for the incident flux. Dark signals are also measured after each influx and efflux measurement.

The sample is placed in an appropriate sample holder, based on the size and shape of the sample. The sample translation stages are adjusted until a 550 nm beam of light from the lamp, with a selected beam diameter, is centered on the sample. The source is then changed to the He-Ne alignment laser at 632.8 nm, and the tilt and rotation of the sample are adjusted until the retro-reflected beam is collinear with the incident beam.

Following alignment of the sample, the appropriate lamp source is selected, turned on, and allowed to equilibrate for a period of one hour to reach stable operation. The appropriate detector and related electronics are selected, and signal levels and stability are checked. The input parameters are selected in the program. They include the source, detector, sample position, wavelength range, and wavelength increment. The measurement is initiated using the scan routine of the software program described in Section 3.4, which can incorporate a list of arbitrary wavelengths. The resulting data files are transferred to appropriate directories.

If the sample being measured is similar to previously measured samples or check standards, then the repeatability will be obtained from a previous measurement, and the sample being calibrated will only be measured once. Repeatability of the instrument is typically determined when the check standards are measured for instrument validation.

Customer-supplied samples are inspected prior to measurement, and the condition is noted in the calibration report. The samples are kept in the original container provided by the customer until the measurements are made. Samples are typically handled by the edges using powder free gloves, and they are not cleaned except for the removal of loose dust using an air bulb blower. Customers are responsible for ensuring the cleanliness quality, surface quality, etc. before submission. Customers who are not satisfied with the condition of a sample should either clean or replace the sample prior to sending it to NIST for calibration.

3.4 Data Acquisition

A Visual Basic program is used for data acquisition and analysis. (Certain commercial equipment or materials are identified in this paper to foster understanding. Such identification does not imply recommendation or endorsement by NIST, nor does it imply that the equipment or materials are necessarily the best available for the purpose.) The computer is connected to the equipment through GPIB and RS-232 interfaces. For each sample measurement, four data files are generated: raw, format, final, and combined. The files are stored in hierarchy-specific directories. All files are saved in the comma-separated-variable (CSV) format.

The raw file (*.raw) stores the multiple digital voltmeter readings of the signal for each sample stage position at a given wavelength and polarization. Each data entry consists of a reading made at one polarization, wavelength, and position with the associated gain and voltmeter electrical range for the detector. The positions correspond to efflux through the sample, influx, or no flux and are referred to as sample, clear, or dark, respectively.

The format file (*.csv) groups the readings from each position for a given wavelength and calculates the signal, or average of all readings, for a given polarization and wavelength. This enables the transmittance to be calculated from the data in each set. The repeatability uncertainty component is the standard deviation of the readings without moving the sample.

The final file (*.cal) presents the values of transmittance for each polarization and the average of both polarizations as a function of wavelength. The calculation procedure is detailed in Section 3.5 below.

The combined file (*.cmb) calculates an average repeat transmittance value based on multiple scans and also calculates the uncertainties for repeatability, wavelength, and linearity. The repeatability uncertainty contribution is evaluated from the standard deviation of repeat measurements, determined in advance as a function of wavelength and absorbance. The wavelength uncertainty contribution is evaluated from the derivative of the transmittance with respect to wavelength. The linearity uncertainty contribution includes effects from both the detector and the signal electronics. See also Section 3.6 for the uncertainty analysis. In most cases the repeatability is determined previously from multiple scans of identical check standards. The combined files contain a compilation of the final values from multiple files when transmittance is measured, and are generated automatically after all the scans are complete. Details of the data analysis and uncertainties are given in Sections 3.5 and 3.6. The data from the combined file are used for the values in the calibration report.

3.5 Data Analysis

The transmittance values in the combine files are initially scanned for outlying data points, due, for instance, to a power line glitch. Any outliers are either substituted by interpolated values or values from other scans of the same sample when available. Interpolated values will be noted in the calibration report. If multiple detectors and/or sources are used, the values from both files are combined and overlap between spectral regions compared.

The data are reduced automatically to transmittance τ in the software files using the following procedures. For each set of n readings at polarization σ, wavelength λ, and position, the average signal $S(\sigma, \lambda)$ is given by

$$S(\lambda,\sigma) = \frac{1}{n}\sum_{i=1}^{n} S(\lambda,\sigma,i). \tag{7}$$

For each polarization and wavelength, the transmittance $\tau(\sigma, \lambda)$ is given by

$$\tau(\lambda,\sigma) = \frac{S_s(\lambda,\sigma) - S_{s,d}(\lambda,\sigma)}{S_c(\lambda,\sigma) - S_{c,d}(\lambda,\sigma)} \cdot \frac{G_c}{G_s}, \tag{8}$$

where G_c and G_s are the gains for the clear and sample signal measurements, respectively. The order of measurement is S_c (clear signal), $S_{c,d}$ (dark clear signal), S_s (sample signal), $S_{s,d}$ (dark sample signal), and S_c. The net signal is the signal with the clear beam path minus the signal with the shutter blocking the beam path. If the lock-in amplifier is used as the electrical detector, no signal is measured with the shutter blocking the beam path, in which case this signal is set to zero. For an unpolarized influx, the spectral transmittance $\tau(\lambda)$ is given by

$$\tau(\lambda) = \frac{1}{2}[\tau(\lambda,s) + \tau(\lambda,p)], \tag{9}$$

where s (perpendicular) and p (parallel) refer to the polarization of the beam with respect to the plane of illumination. This equation is equivalent to Eq. (1).

Reducing the data to final values for multiple source and detector combinations is usually accomplished using the combined files and a spreadsheet program. In general, each sample is measured once over a specified wavelength range, with the appropriate sources and detectors. If the sample is scanned more than once, the final values in a calibration report are obtained in a two-step process. First, the values measured multiple times under the same conditions are averaged in the combined file. Second, the final values for all source and detector combinations are plotted as a function of wavelength.

3.6 Uncertainties

The uncertainties for the calibration reports are calculated according to the procedures outlined in "Guidelines for Evaluating and Expressing the Uncertainty of NIST Measurement Results" [7]. There are several ways to classify the uncertainties in the transmittance values. Uncertainty can arise from systematic or random effects; the evaluation of the uncertainty can be classified as Type A or B; and the source of uncertainty can be sample-dependent or sample-independent. The sources of uncertainty for the transmittance measurements are given in Table 6, along with their classification due to effect and their dependence on the properties of the sample. Repeatability is defined as the closeness of agreement between results of successive measurements of a sample using the same parameters. The repeatability of the transmittance of a sample can be obtained by simply starting another spectral scan at another point in time. An example of a factor that could affect the repeatability of a measurement is the stability of the lamp. In contrast, reproducibility is defined as the closeness of agreement between results for measurements of a sample using the same parameters, but under changed conditions. For example, the reproducibility of the transmittance of a sample can be obtained by removing the sample from the sample holder, restarting the lamp, and performing the measurement on different days. Sample uniformity and positioning can affect the reproducibility of a measurement.

Table 6. Sources of uncertainty and classifications for transmittance measurements
(S: Systematic; R: Random; I: Independent; D: Dependent)

Source of Uncertainty	Type of Effect	Uncertainty Type	Sample Dependence
Wavelength	S	B	D
Stray Light	S	B	D
Detector Linearity (DL)	S	A	D
Detector Range Ratio (DRR)	S	B	I
Detector Gain Ratio (DGR)	S	B	I
Repeatability (RPT)	R	A	D
Reproducibility (RPD)	S	B	D

Note that most of the sources of uncertainty arise from systematic effects.
The total relative uncertainty in the measured transmittance is given by the expression,

$$\frac{u(\tau)}{\tau} = \left\{ \left[\frac{u_0(\tau)}{\tau}\right]^2 + \left[\frac{u(\lambda)}{\lambda}\right]^2 + \left[\frac{u(\mathrm{DL})}{\mathrm{DL}}\right]^2 + \left[\frac{u(\mathrm{DRR})}{\mathrm{DRR}}\right]^2 + \left[\frac{u(\mathrm{DGR})}{\mathrm{DGR}}\right]^2 \right.$$
$$\left. + \left[\frac{u(\mathrm{RPT})}{\mathrm{RPT}}\right]^2 + \left[\frac{u(\mathrm{RPD})}{\mathrm{RPD}}\right]^2 \right\}^{1/2} , \qquad (10)$$

where DL, DRR, DGR, RPT, and RPD are defined in the first column of Table 6 and

$$u_0(\tau) = \left\{ \left[\frac{\partial \tau}{\partial S_s} \cdot u(S_s)\right]^2 + \left[\frac{\partial \tau}{\partial S_{s,d}} \cdot u(S_{s,d})\right]^2 + \left[\frac{\partial \tau}{\partial S_c} \cdot u(S_c)\right]^2 + \left[\frac{\partial \tau}{\partial S_{c,d}} \cdot u(S_{c,d})\right]^2 \right.$$
$$\left. + \left[\frac{\partial \tau}{\partial G_s} \cdot u(G_s)\right]^2 + \left[\frac{\partial \tau}{\partial G_c} \cdot u(G_c)\right]^2 \right\}^{1/2} . \qquad (11)$$

In Eq. (10), the uncertainty contribution caused by wavelength, $u(\lambda)/\lambda$, is evaluated by using the expression,

$$\frac{u_s(\lambda)}{\lambda} = \frac{d\tau}{d\lambda} \cdot \frac{u_i(\lambda)}{\tau} , \qquad (12)$$

where $u_s(\lambda)$ is the standard sample uncertainty in wavelength, $u_i(\lambda)$ is the standard instrument uncertainty in wavelength, and the derivative of the transmittance is calculated from a cubic spline fit of transmittance as a function of wavelength. For the 600 groove/mm gratings, the standard instrument uncertainties in wavelength $u_i(\lambda)$ are 0.05 nm for the one blazed at 200 nm (position A) and 0.10 nm for the one blazed at 1250 nm (position B).

The sources of uncertainty and their uncertainty contributions, expressed in absolute terms, are provided in the calibration report. The sources of uncertainty due to wavelength and repeatability can be wavelength dependent, which can result in different uncertainty contributions for different wavelengths or wavelength ranges. In cases where the same detector range or gain setting is used for both the clear and sample signals, there is no uncertainty caused by detector range or gain ratios and these sources are not listed in the report. The uncertainty contribution due to repeatability is evaluated from the propagated uncertainties in the signals or from the standard deviation of repeat measurements, and is usually averaged over wavelength ranges. Finally, the expanded uncertainty is the combined (root-sum-square) uncertainty from all the uncertainty contributions, multiplied by a coverage factor $k = 2$. Typical uncertainty contributions for transmittance measured using a Si photodiode are given in Table 7, where percent signs refer to a percentage of the current value of the transmittance.

Table 7. Sources of uncertainty and typical uncertainty contributions at 0.5 transmittance

Source of Uncertainty	Standard Uncertainty	Uncertainty Contribution
Wavelength (λ)	0.05 nm	0.0001
Detector Linearity (DL)	0.05 %	0.0003
Detector Range Ratio (DRR)	0.02 %	0.0001
Detector Gain Ratio (DGR)	0.1 %	0.0005
Repeatability (RPT)	0.02 %	0.0001
Reproducibility (RPD)	0.035 %	0.0002
Expanded Uncertainty ($k = 2$)		0.0012

The sources of uncertainties for various detector systems used in transmittance measurements are listed in Table 8.

Table 8. Sources of uncertainties for various detector systems

Detector[a]		Source of Uncertainty	Gain	Uncertainty
Optical	Electrical			
PMT	DVM	Linearity		0.2 %
		Range Ratio		0.02 %
		Reproducibility		0.03 %
Si	DVM	Linearity		0.05 %
		Range Ratio		0.02 %
		Gain Ratio	7-8	0.01 %
			8-9	0.1 %
			7-9	0.1 %
		Reproducibility		0.035 %
Si	LIA	Linearity		0.05 %
		Range Ratio		0.02 %
		Reproducibility		0.035 %
Extended InGaAs	LIA	Linearity		0.075 %
		Range Ratio		0.1 %[a]
		Reproducibility		0.03 %

[a] PMT – photomultiplier tube; digital voltmeters DVM – digital voltmeter; Si – silicon photodiode; and LIA – lock-in amplifier

3.7 Calibration Report

The calibration report contains a description of the sample and the test performed. Also included are the data from the measurements and the associated uncertainties. Measurements of room temperature and sample dimensions are provided as reference and are not intended to be certified values. A plot of the data is also provided. The scale displayed on the plot may not be the best representation for the customer's application.

The calibration sample and a photocopy of the report are mailed together. Upon request, the report and the tabulated data in unencrypted form may be provided to the customer by email. The method of shipping will be determined by NIST unless the customer provides shipping account information and instruction. NIST does not provide any insurance above the free insurance offered by the shipping agent. Special arrangement can be made to pick up items in person. The original report is mailed separately from the sample. NIST maintains a file on the calibration for a period of three years.

4. Transfer Instrument

The transfer instrument is used for the majority of calibrations that require extended wavelength ranges or sets of multiple filters. The transfer instrument used is a commercial spectrophotometer that performs similar measurements to the RTS with less setup time and more rapid scan rates. It is a dual-beam instrument with a double monochromator, and it can operate in either a regular transmittance or diffuse reflectance mode. The uncertainties are larger than those of the RTS, but provide services at a considerable cost savings to the customer.

While the RTS establishes the national scale for regular spectral transmittance by providing the highest accuracy, it requires significant time to set up and scan samples over large spectral regions. Therefore the RTS is typically reserved for use in establishing the scale using samples that can transfer the scale to transfer instruments.

4.1 Instrument Description

The overall wavelength range of the instrument is 185 nm to 3150 nm. Based on the transfer of the transmittance scale from the reference instrument, the calibration range is limited to wavelengths from 350 nm to 2500 nm. The monochromator uses two gratings, one to cover the UV-Vis (350 nm to 900 nm) range, and the other covering the NIR (700 nm to 2500 nm) range. A QTH lamp is used as the source. The scan rate can be selected between 0.004 nm/s and 2000 nm/s.

The spectral bandwidth can be set to any fixed value between 0.01 nm to 5.00 nm in the UV-Vis, and 0.04 nm to 20.0 nm in the NIR. The minimum bandwidth may be limited depending on the available signal.

The photometric range in the transmittance mode is 0 % to 100 %, while the photometric range in the transmittance density is 0 to 4.5. The transmittance density is defined as the common logarithm (base ten) of the reciprocal transmittance,

$$D = -\log_{10} \tau \ . \tag{13}$$

The transmittance measurements are performed at ambient temperature and humidity. The temperature is noted at the time of the measurement. The allowable operating range of the instrument for relative humidity is 20 % to 80 %, based on the manufacturer's specifications.

The instrument is connected to the computer via GPIB. The computer is used to set the measurement parameters, and to display, collect, and analyze the data. Files of customer measurements are stored on a centralized server.

4.2 Instrument Calibration

Validation of the instrument operation is achieved by performing tests with the validation programs which are part of the instrument's software package. These tests are run every year or before a critical measurement, such as a calibration that does not rely on master standards.

The validation tests rely upon various standards. The wavelength scale is validated by scans of deuterium (D_2) and mercury (Hg) lamps, which are internal to the instrument, and by a holmium oxide (Ho_2O_3) glass, which uses the master sample. The wavelengths of minimum transmittance for the master Ho_2O_3 sample are given in Table 9. The photometric scale is validated by scans of neutral-density glass filters, SRM 930 and SRM 1930. The regular spectral transmittance of these filters is determined quarterly on the RTS. A 2" x 2" sample holder or a cuvette holder can be used for the transfer standards.

Table 9. Certified and measured wavelengths of minimum transmittance of a holmium oxide glass filter, for a spectral bandwidth of 1 nm [8, 9]

Band	Certified wavelengths [nm]
1	241.6
2	279.3
3	287.6
4	333.9
5	360.9
6	386.0
7	418.8
8	453.6
9	460.2
10	536.5
11	637.8

4.3 Measurement Procedure

Samples are cleaned with an air bulb to remove loosely attached dust particles only. They are kept in the container provided by the customer until the measurement is made, and are typically handled by the edges using powder-free gloves.

The instrument is turned on and allowed to warm up for a period of one hour. The measurement parameters are input into the instrument software. These include the wavelength range, data interval, averaging time, spectral band width, and source and detector change-over.

A 100 % transmittance baseline is obtained by scanning a blank sample holder. The blank sample holder is blocked with an opaque material for the 0 % transmittance baseline. A zero-baseline correction can either be calculated automatically by the software or manually by the user.

4.4 Data Analysis

The software accompanying the instrument reduces the data to percent transmittance and is saved in a CSV format. The data are plotted and analyzed for errors. Once the data have been analyzed, the data are formatted into the calibration report in the same manner as done for the RTS results.

4.5 Uncertainties

The uncertainties for transmittance calibrations on the transfer instrument are based on the uncertainties from the RTS. The scale of regular spectral transmittance is transferred from the RTS using check standards as transfer standards. The difference measured for the same sample between the two instruments is used as the estimate for the uncertainty $u(\tau_{r-t})$. The total expanded uncertainty for transmittance $u(\tau_t)$ as measured on the transfer instrument is calculated from the following expression,

$$u(\tau_t) = \sqrt{[u(\tau_r)]^2 + [u(\tau_{r-t})]^2}, \qquad (14)$$

where $u(\tau_r)$ is the transmittance uncertainty for the RTS from Eq. (10).

Acknowledgments

The authors gratefully acknowledge Gerald Fraser, John Travis, Eric Shirley, and Robert Saunders for careful reading of this manuscript.

References

1. P. Y. Barnes, E. A. Early, and A. C. Parr, "Spectral Reflectance," *NIST Spec. Pub.* **250-48**, U. S. Dept. of Commerce (1998).
2. M. E. Nadal, E. A. Early, and E. A. Thompson, "Specular Gloss," *NIST Spec. Pub.* **1026**, U. S. Dept. of Commerce (2006).
3. K. L. Eckerle, J. J. Hsia, K. D. Mielenz, and V. R. Weidner, "NBS Measurement Services: Regular Spectral Transmittance," *NBS Spec. Pub.* **250-6**, U. S. Dept. of Commerce (1987).
4. D. W. Allen, "Holmium Oxide Glass Wavelength Standards," *J. Res. Natl. Inst. Stand. Technol.*, **112**, pp. 303-306 (2007).
5. J. E. Stewart and W. S. Gallaway, "Diffraction Anomalies in Grating Spectrometers," *Applied Optics*, **1**, 421 (1962).
6. K. D. Mielenz and K. L. Eckerle, "Spectrophotometer Linearity Testing using Double Aperture Method," *Applied Optics*, **11**, pp. 2295-2303 (1972).
7. B. N. Taylor, and C. E. Kuyatt, "Guidelines for evaluating and expressing uncertainty on NIST measurement results," *NIST Technical Note*, **1297** (1994).
8. H. J. Keegan, J. C. Schleter, and V. R. Weidner, "Ultraviolet Wavelength Standard for Spectrophotometry," *J. Opt. Soc. Am.*, **51**, 1470 (1961).
9. "Standards for Checking the Calibration of Spectrophotometers (200 to 1000 nm)," Letter Circular 1017, U. S. Dept. of Commerce (1967).
10. K. D. Mielenz and K. L. Eckerle, "Design, Construction, and Testing of a New High Accuracy Spectrophotometer," *NBS Tech. Note* **729**, U. S. Dept. of Commerce (1972).
11. K. L. Eckerle, "Modification of an NBS Reference Spectrophotometer," *NBS Tech. Note* **913**, U. S. Dept. of Commerce (1976).
12. IUPAC Compendium of Chemical Terminology, http://www.chemsoc.org/chembytes/goldbook/index.htm.
13. "Standard Practice for Computing the Colors of Objects by Using the CIE System, ASTM E308-95," American Society for Testing and Materials, West Conshohocken, PA (1995).
14. Z.M. Zhang, R.R. Gentile, A.L. Migdall, and R.U. Datla, "Transmittance Measurements for Filters of Optical Density between One and Ten," *Applied Optics*, **36**, pp. 8889-8895 (1997).
15. K. L. Eckerle, J. J. Hsia, and V. R. Weidner, "Transmittance MAP Service," NIST SP692, U.S. Government Printing Office, Washington, D.C. (1985).

Appendix A: Transmittance Theory and Measurement Equation

When radiant flux (optical radiation) interacts with matter, it may be reflected, absorbed, or transmitted. For purposes of the regular spectral transmittance measurement service, the following definitions [10, 11] apply.

Transmission is the process whereby radiant flux passes through a material or object without a change in wavelength of its monochromatic components. Therefore, photoluminescence – luminescence produced by the absorption of radiant flux – is not considered here. Diffusion is a change of the angular distribution of a beam of radiant flux by a transmitting material or a reflecting surface, such that flux incident in one direction is continuously distributed in many directions. The process does not conform on a macroscopic scale to the laws of Fresnel reflection and refraction and does not change the wavelength of the monochromatic components of the flux. Regular transmission is transmission in accordance with the laws of geometrical optics, without diffusion, while diffuse transmission is transmission in which diffusion occurs, independently of the laws of refraction. Mixed transmission is a combination of regular and diffuse transmission.

The transmittance τ of an object is the ratio of the transmitted radiant flux Φ_t to the incident radiant flux Φ_i, under specified geometrical and spectral conditions, and is given by

$$\tau = \frac{\Phi_t}{\Phi_i} \ . \tag{A1}$$

Transmittance depends on wavelength, polarization, and on the angles of incidence and collection. The internal transmittance τ_i of an object is the ratio of the radiant flux reaching the exit surface to the radiant flux that penetrates the entry surface. The reflectance ρ of an object is the ratio of the reflected radiant flux Φ_r to the incident radiant flux, under specified geometric and spectral conditions, and is given by

$$\rho = \frac{\Phi_r}{\Phi_i} \ . \tag{A2}$$

The absorptance α of an object is the ratio of the absorbed radiant flux Φ_a to the incident radiant flux, under specified geometric and spectral conditions, and is given by

$$\alpha = \frac{\Phi_a}{\Phi_i} \ . \tag{A3}$$

By conservation of energy, the transmittance, reflectance, and absorptance sum to one. That is,

$$\tau + \rho + \alpha = 1 \ . \tag{A4}$$

Relating the measured transmittance of an object, which is an extrinsic property, to its intrinsic properties requires consideration of both the interaction of the radiant flux with the boundary of the object and with the material composing the interior. Since in nearly all cases for which regular transmittance is measured, the object is semi-transparent and non-conducting, the following equations are applicable for a homogeneous dielectric, linear, isotropic, and non-magnetic medium.

For the boundary interaction, a ray of radiant flux is incident upon the boundary of an object, as shown in Fig. A1. The angles between the normal of the boundary and the incident, reflected, and transmitted rays are θ_i, θ_r, and θ_t, respectively. The material from which the ray is incident has index of refraction n_i, while the material into which the ray is transmitted has index of refraction n_t. The plane of incidence is the plane containing the surface normal and the incident ray.

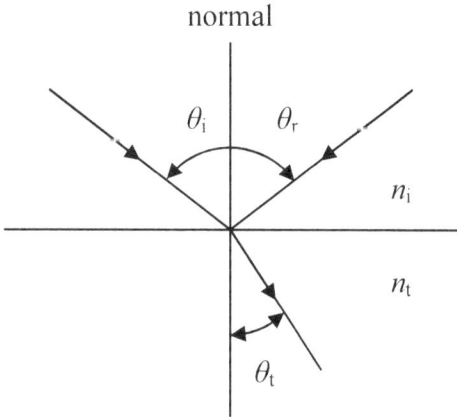

Figure A1. Interactions of incident radiation with a surface boundary.

The law of reflection states that the reflected ray is in the plane of incidence and the angles are related by

$$\theta_r = \theta_i \,. \tag{A5}$$

The law of refraction, or Snell's Law, states that the transmitted ray is also in the plane of incidence and the angles are related by

$$n_t \sin \theta_t = n_i \sin \theta_i \,. \tag{A6}$$

The Fresnel equations for the reflection and transmission of the ray at the boundary are obtained from Maxwell's equations and the boundary conditions for the electric and magnetic fields at the boundary. These equations depend on the polarization σ of the incident ray, with s and p polarization having the electric field perpendicular and parallel to the plane of incidence, respectively. The amplitude reflection coefficients r depend on the angle of incidence, the wavelength λ, and the polarization, and are given by

$$r(\theta_i, \lambda, \text{s}) = \frac{n_i(\lambda)\cos\theta_i - n_t(\lambda)\cos\theta_t}{n_i(\lambda)\cos\theta_i + n_t(\lambda)\cos\theta_t} \text{ and} \tag{A7}$$

$$r(\theta_i, \lambda, \text{p}) = \frac{n_t(\lambda)\cos\theta_i - n_i(\lambda)\cos\theta_t}{n_t(\lambda)\cos\theta_i + n_i(\lambda)\cos\theta_t} . \tag{A8}$$

The angle of the transmitted ray can be eliminated from Eqs. (A7) and (A8) by using Eq. (A6). Similarly, the amplitude transmission coefficients t are given by

$$t(\theta_i, \lambda, \text{s}) = \frac{2n_i(\lambda)\cos\theta_i}{n_i(\lambda)\cos\theta_i + n_t(\lambda)\cos\theta_t} \text{ and} \tag{A9}$$

$$t(\theta_i, \lambda, \text{p}) = \frac{2n_i(\lambda)\cos\theta_i}{n_t(\lambda)\cos\theta_i + n_i(\lambda)\cos\theta_t} . \tag{A10}$$

The radiant flux of a ray is given by the irradiance, the magnitude of the Poynting vector, multiplied by the illumination source area projected onto the sample plane. The reflectance and transmittance at the boundary are the ratio of the reflected and transmitted radiant flux, respectively, to the incident radiant flux. In terms of the amplitude coefficients, the boundary reflectance ρ_b is given by

$$\rho_b(\theta_i, \lambda, \sigma) = r^2(\theta_i, \lambda, \sigma) , \tag{A11}$$

and the boundary transmittance τ_b is given by

$$\tau_b(\theta_i, \lambda, \sigma) = \frac{n_t \cos\theta_t}{n_i \cos\theta_i} t^2(\theta_i, \lambda, \sigma) . \tag{A12}$$

Within the object, the radiant flux interacts with the medium. According to Bouguer's law, the internal transmittance, an extrinsic property of the object, is given by

$$\tau_i(\lambda, d) = e^{-\mu(\lambda)d} , \tag{A13}$$

where $\mu(\lambda)$ is the spectral attenuation coefficient, an intrinsic property of the medium, and d is the distance through the object, *i.e.*, thickness of object. The attenuation coefficient is the sum of the absorption coefficient $a(\lambda)$ and the scattering coefficient $s(\lambda)$. The mean penetration distance $d_m(\lambda)$, an intrinsic property of the object, is given by

$$d_m(\lambda) = \frac{1}{\mu(\lambda)} , \tag{A14}$$

while the optical thickness β, an extrinsic property of the object, is given by

$$\beta(\lambda) = \frac{d(\lambda)}{d_m(\lambda)} \,. \tag{A15}$$

Expressing the spectral attenuation coefficient as a product of the concentration c of the absorbing medium and the medium's extinction coefficient $\varepsilon(\lambda)$ yields Beer's law form of Eq. (12), given by

$$\tau_i(\lambda, d) = e^{-c\varepsilon(\lambda)d} \,. \tag{A16}$$

Sometimes, absorbance measurements are requested by the customer. The transmission (optical) density $D(\lambda)$ is given by

$$D(\lambda) = -\log_{10} \tau(\lambda) \,, \tag{A17}$$

while the absorbance $A_i(\lambda)$, or just optical density, is given by

$$A_i(\lambda) = -\log_{10} \tau_i(\lambda) \,. \tag{A18}$$

Note that the transmission density depends on the total transmittance from Eq. (A1), whereas the optical density is a function of the internal transmittance from Eq. (A13). Internal transmittance refers to energy loss by absorption only, whereas the total transmittance is that due to absorption, reflection, scatter, and other processes [12]. Conversely, the transmittance calculated from the absorbance is given by

$$\tau_i(\lambda) = 10^{-A_i(\lambda)} \,. \tag{A19}$$

Suppressing λ-dependence from Eq. (A18), the uncertainty in absorbance, $u(A)$, in terms of $u(\tau)$ is given by

$$u(A) = -0.434 \frac{u(\tau)}{\tau} \,. \tag{A20}$$

Likewise from Eq. (A19), the uncertainty in transmittance, $u(\tau)$, in terms of $u(A)$ is given by

$$u(\tau) = -2.303 \cdot 10^{-A} \cdot u(A) \,. \tag{A21}$$

For two absorbances A_1 and A_2 corresponding to two thickness d_1 and d_2 of the same medium, Lambert's law is given by

$$A_1 = \frac{d_1}{d_2} A_2 \,. \tag{A22}$$

Although the luminous transmittance is not often requested, sometimes it is needed. The luminous transmittance τ_V for CIE standard illuminant C and for both the CIE 1931 (2°) standard observer color-matching function $\bar{y}$ and the CIE 1964 (10°) supplemental standard color-matching function $\bar{y}_{10}$ is given by

$$\tau_V = \frac{\sum_\lambda S(\lambda) \cdot \tau(\lambda) \cdot \bar{y}(\lambda) \cdot \Delta\lambda}{\sum_\lambda S(\lambda) \cdot \bar{y}(\lambda) \cdot \Delta\lambda} \, , \tag{A23}$$

where λ is the wavelength, $S(\lambda)$ is the spectral power distribution of CIE standard illuminant C, $\tau(\lambda)$ is the regular spectral transmittance of the calibration item, $\bar{y}(\lambda)$ is the appropriate color-matching function, and $\Delta\lambda$ is the spectral bandwidth. Values for $S(\lambda)$ and $\bar{y}(\lambda)$ were obtained from the ASTM E308 documentary standard [13].

An expression for the measured transmittance of an object of finite thickness in terms of the properties of the boundary and medium is obtained by considering the inter-reflections as radiant flux passes through the object in Fig. A2. If a ray is incident upon the boundary of the object, part of the ray is reflected by the boundary, while the remainder passes into the interior. The ray is attenuated as it passes through the object, and is both reflected and transmitted by the other boundary. The fraction of the ray that is reflected is further attenuated as it passes through the object to the original boundary. This process of reflection, transmission, and attenuation is repeated, and the total radiant flux that is transmitted through the object is measured. The measured absorptance, in terms of the boundary reflectance and transmittance and internal transmittance is given by

$$\begin{aligned} \alpha(\theta_i, \lambda, \sigma, d) &= (1 - \tau_i) \cdot \tau_b \cdot \left[1 + \tau_i \cdot \rho_b + (\tau_i \cdot \rho_b)^2 + (\tau_i \cdot \rho_b)^3 + \cdots \right] \\ &= \frac{(1 - \tau_i) \cdot \tau_b}{1 - \tau_i \cdot \rho_b} \end{aligned}, \tag{A24}$$

where the dependences on incident angle, wavelength, polarization, and thickness on the right-hand-side of the equation have been suppressed for clarity. Likewise, the measured reflectance is given by

$$\begin{aligned} \rho(\theta_i, \lambda, \sigma, d) &= \rho_b + \tau_i^2 \cdot \tau_b^2 \cdot \rho_b \cdot \left[1 + (\tau_i \cdot \rho_b)^2 + (\tau_i \cdot \rho_b)^4 + (\tau_i \cdot \rho_b)^6 + \cdots \right] \\ &= \rho_b + \frac{\tau_i^2 \cdot \tau_b^2 \cdot \rho_b}{1 - (\tau_i \cdot \rho_b)^2} = \rho_b \left[1 + \frac{\tau_i^2 \cdot \tau_b^2}{1 - (\tau_i \cdot \rho_b)^2}\right] \end{aligned} \tag{A25}$$

and the measured transmittance is given by

$$\begin{aligned} \tau(\theta_i, \lambda, \sigma, d) &= \tau_i \cdot \tau_b^2 \cdot \left[1 + (\tau_i \cdot \rho_b)^2 + (\tau_i \cdot \rho_b)^4 + (\tau_i \cdot \rho_b)^6 + \cdots \right] \\ &= \frac{\tau_i \cdot \tau_b^2}{1 - (\tau_i \cdot \rho_b)^2} \end{aligned}. \tag{A26}$$

The above formulas neglect coherence effects important when the coherence length of the incident light,

$$L \cong \frac{\lambda^2}{n(\Delta\lambda)}, \qquad (A27)$$

is on the order of or larger than the sample thickness. For $\lambda = 1000$ nm, $\Delta\lambda = 1.5$ nm, and $n = 1.5$, the coherence length is 0.44 mm. Zhang et al. discusses the effects of complete and partial coherence for transmittance measurements of optical filter using a Nd:YAG laser. [14]

The sum of Eqs. (A24) to (A26) is one, as required by the conservation of energy. The application of Eqs. (A24) to (A26) is illustrated by several examples. First, consider two extreme cases. For an opaque object, $\tau_i = 0$, which yields $\alpha = 1 - \rho_b = \tau_b$, $\rho = \rho_b$, and $\tau = 0$. For a perfectly transparent object, $\tau_i = 1$, which yields $\alpha = 0$,

$$\rho = \frac{2\rho_b}{1+\rho_b}, \text{ and} \qquad (A28)$$

$$\tau = \frac{1-\rho_b}{1+\rho_b}. \qquad (A29)$$

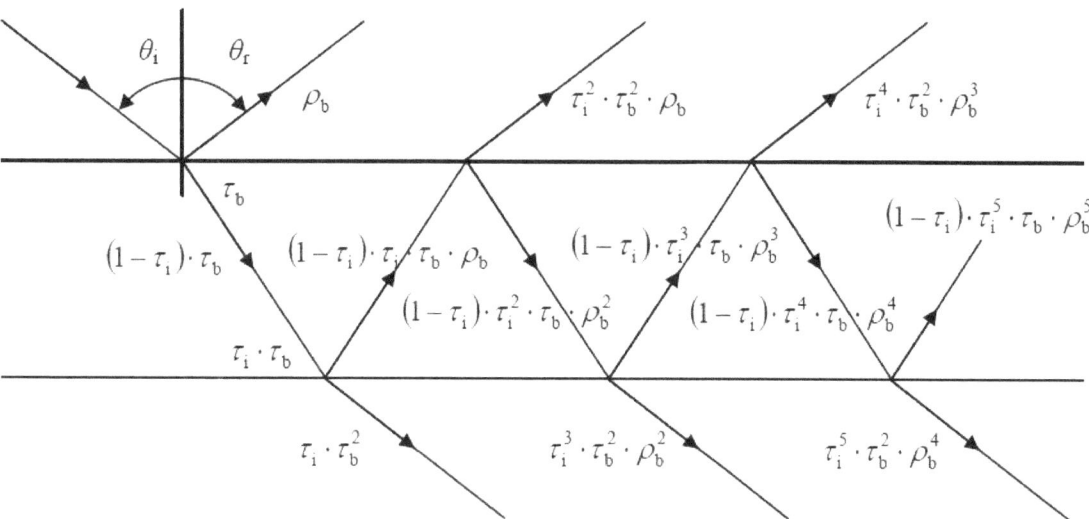

Figure A2. Interactions of incident radiation within a layer.

For a more realistic case, consider an object composed of glass with a thickness of 2 mm. The measurement is performed in air, so $n_i = 1$, and the index of refraction of the glass is $n_t = 1.5$. The attenuation coefficient is $\mu = 0.005$ mm^{-1}, yielding an internal transmittance at normal incidence of $\tau_i = 0.99005$, a mean penetration distance of $d_m = 200$ mm, and an optical thickness of $\beta = 0.01$. For a given angle of incidence, the path length is given by

$$d = \frac{2 \text{ mm}}{\cos \theta_t} \,. \tag{A30}$$

The transmittance as a function of incident angle for s and p polarizations and for unpolarized incident radiant flux is shown in Fig. A3.

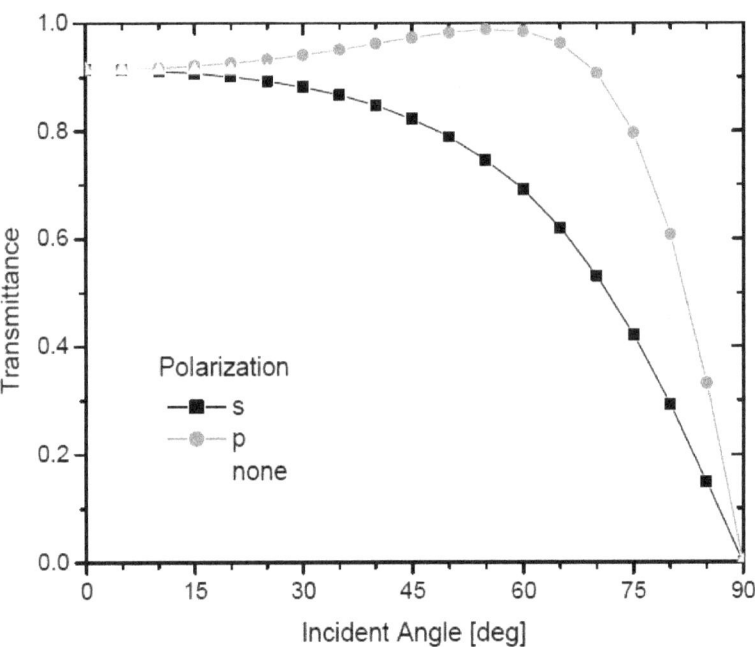

Figure A3. Transmittance as a function of incident angle for the indicated polarizations of the incident radiant flux of a sample with an index of refraction of 1.5, thickness of 2 mm, and attenuation coefficient of 0.005 mm^{-1}.

The Reference Transmittance Spectrophotometer (RTS) typically measures the regular transmittance of a sample at normal incidence ($\theta_i = 0°$) as a function of wavelength and polarization of the illumination beam. Measuring at normal incidence eliminates effects from polarization and path length through the sample. This appendix derives the measurement equation corresponding to the operation of the RTS, which relates measured quantities to the transmittance.

As detailed in Section 3.1, the RTS operates by polarizing a collimated, monochromatic illumination beam and passing it through the sample carriage. The beam continues on to the receiver, whose optical detector measures a signal proportional to the radiant flux of the beam. The sample carriage consists of two incident positions for the beam: clear and sample. In each position a shutter is used as a light trap for a dark signal. At each wavelength and polarization, signals are measured from the clear, dark, sample, dark, and clear positions, in order. Signals for the clear and sample positions are proportional to the incident and transmitted radiant fluxes, respectively. Net signals for the clear and sample positions are obtained by subtracting the dark position signals. Since the beam emerging from the collimating mirror is not completely polarized, measurements are made for two orthogonal polarizations of the illumination beam, and the resulting transmittances are averaged to obtain the transmittance for unpolarized incident radiant flux. For typical samples measured so far, the variation due to polarization is within the expanded uncertainty.

To derive the measurement equations, several definitions are needed. Since the illumination beam of the RTS is linearly polarized, all of the following polarizations of the radiant flux are assumed to be along either the s or p direction. Expanding upon the definition of transmittance given by Eq. (A1), the transmittance for polarized incident and transmitted radiant flux is given by

$$\tau(\lambda, \sigma_i, \sigma_t) = \frac{\Phi_t(\lambda, \sigma_i, \sigma_t)}{\Phi_i(\lambda, \sigma_i)} , \qquad (A31)$$

where λ is the wavelength of the radiant flux, and σ_i and σ_t are the polarization of the incident and transmitted radiant flux, respectively. Note that the polarization of the transmitted radiant flux can be different from that of the incident radiant flux. In most cases, the total transmitted radiant flux for a given polarization of the incident radiant flux is required, given by

$$\tau(\lambda, \sigma_i, t) = \frac{\Phi_t(\lambda, \sigma_i, t)}{\Phi_i(\lambda, \sigma_i)} = \frac{\Phi_t(\lambda, \sigma_i, s) + \Phi_t(\lambda, \sigma_i, p)}{\Phi_i(\lambda, \sigma_i)} , \qquad (A32)$$

where t denotes the total transmitted radiant flux. The transmittance for unpolarized incident radiant flux and polarized incident radiant flux is also useful, and is given by

$$\tau(\lambda, u, \sigma_t) = \frac{\Phi_t(\lambda, u, \sigma_t)}{\Phi_i(\lambda, u)} = \frac{\Phi_t(\lambda, s, \sigma_t) + \Phi_t(\lambda, p, \sigma_t)}{\Phi_i(\lambda, s) + \Phi_i(\lambda, p)} , \qquad (A33)$$

where u denotes unpolarized incident radiant flux. Note that the incident radiant fluxes for each polarization have to be equal. Finally, the transmittance for unpolarized incident radiant flux and the total transmitted radiant flux is given by

$$\tau(\lambda, u, t) = \frac{\Phi_t(\lambda, u, t)}{\Phi_i(\lambda, u)} = \frac{\Phi_t(\lambda, s, s) + \Phi_t(\lambda, s, p) + \Phi_t(\lambda, p, s) + \Phi_t(\lambda, p, p)}{\Phi_i(\lambda, s) + \Phi_i(\lambda, p)} . \qquad (A34)$$

The radiant flux is not measured directly. Rather, a current I [A] from the optical detector is measured, which is proportional to the radiant flux Φ [W] by

$$I(\lambda, \sigma) = \Phi(\lambda, \sigma) \cdot R(\lambda, \sigma) , \qquad (A35)$$

where R [A/W] is the responsivity of the detector, which depends on the wavelength and polarization of the radiant flux. Following the progression for transmittance given by Eqs. (A31) to (A35), the currents corresponding to polarized, total, and unpolarized incident and transmitted radiant fluxes are given by

$$I_i(\lambda, \sigma_i) = \Phi_i(\lambda, \sigma_i) \cdot R(\lambda, \sigma_i) \, , \tag{A36}$$

$$I_t(\lambda, \sigma_i, \sigma_t) = \Phi_t(\lambda, \sigma_i, \sigma_t) \cdot R(\lambda, \sigma_t) \, , \tag{A37}$$

$$I_t(\lambda, \sigma_i, t) = \Phi_t(\lambda, \sigma_i, s) \cdot R(\lambda, s) + \Phi_t(\lambda, \sigma_i, p) \cdot R(\lambda, p) \, , \tag{A38}$$

$$I_i(\lambda, u) = \Phi_i(\lambda, s) \cdot R(\lambda, s) + \Phi_i(\lambda, p) \cdot R(\lambda, p) \, , \tag{A39}$$

and

$$I_t(\lambda, u, t) = \Phi_t(\lambda, s, s) \cdot R(\lambda, s) + \Phi_t(\lambda, s, p) \cdot R(\lambda, p) + \Phi_t(\lambda, p, s) \cdot R(\lambda, s) + \Phi_t(\lambda, p, p) \cdot R(\lambda, p) \, . \tag{A40}$$

The transmittances given in Eqs. (A31) to (A34) can now be expressed in terms of the measured currents and known responsivities. Solving Eqs. (A36) and (A37) for the radiant flux and substituting into Eq. (A31) yields

$$\tau(\lambda, \sigma_i, \sigma_t) = \frac{\dfrac{I_t(\lambda, \sigma_i, \sigma_t)}{R(\lambda, \sigma_t)}}{\dfrac{I_i(\lambda, \sigma_i)}{R(\lambda, \sigma_i)}} = \frac{I_t(\lambda, \sigma_i, \sigma_t)}{I_i(\lambda, \sigma_i)} \cdot \frac{R(\lambda, \sigma_i)}{R(\lambda, \sigma_t)} \, . \tag{A41}$$

Calculating $\tau(\lambda, \sigma_i, \sigma_t)$ from the measured currents is therefore possible only if the receiver is able to discriminate the polarization of the radiant flux, by having a polarizer between the sample and the detector, and knowing the ratio of the responsivities for the two polarizations, or by having no change in polarization of the radiant flux that is transmitted by the sample.

The transmittance for the total transmitted radiant flux, from Eqs. (A32), (A36), and (A38), is

$$\tau(\lambda, \sigma_i, t) = \frac{I_t(\lambda, \sigma_i, s)}{I_i(\lambda, \sigma_i)} \cdot \frac{R(\lambda, \sigma_i)}{R(\lambda, s)} + \frac{I_t(\lambda, \sigma_i, p)}{I_i(\lambda, \sigma_i)} \cdot \frac{R(\lambda, \sigma_i)}{R(\lambda, p)} \, . \tag{A42}$$
$$= \tau(\lambda, \sigma_i, s) + \tau(\lambda, \sigma_i, p)$$

This expression for $\tau(\lambda, \sigma_i, t)$ is valid if the receiver is able to discriminate the polarization of the radiant flux. If there is no change in the polarization of the radiant flux that is transmitted by the sample, then Eq. (A42) reduces to Eq. (A41) without the ratio of the responsivities. Finally, if the receiver is not sensitive to the polarization of the radiant flux, we have

$$R(\lambda, s) = R(\lambda, p) \, , \tag{A43}$$

and Eq. (A42) becomes

$$\tau(\lambda,\sigma_i,t) = \frac{I_t(\lambda,\sigma_i,t)}{I_i(\lambda,\sigma_i)}, \tag{A44}$$

where the current from the optical detector $I_t(\lambda, \sigma_i, t)$ includes both polarizations of the transmitted radiant flux.

The transmittance for unpolarized incident radiant flux, from Eqs. (A33), (A37), and (A39) is

$$\tau(\lambda,u,\sigma_t) = \frac{\dfrac{I_t(\lambda,s,\sigma_t)}{R(\lambda,\sigma_t)} + \dfrac{I_t(\lambda,p,\sigma_t)}{R(\lambda,\sigma_t)}}{\dfrac{I_i(\lambda,s)}{R(\lambda,s)} + \dfrac{I_i(\lambda,p)}{R(\lambda,p)}}. \tag{A45}$$

To make the incident radiant fluxes for both s and p polarizations equal, the terms in Eq. (A45) for the p polarization of the incident radiant flux are multiplied by a factor of $(I_i(\lambda, s)/R(\lambda, s))/(I_i(\lambda, p)/R(\lambda, p))$. Multiplying both the numerator and denominator by $R(\lambda, s)/I_i(\lambda, s)$ then yields

$$\begin{aligned}\tau(\lambda,u,\sigma_t) &= \frac{1}{2}\left[\frac{I_t(\lambda,s,\sigma_t)}{I_i(\lambda,s)}\cdot\frac{R(\lambda,s)}{R(\lambda,\sigma_t)} + \frac{I_t(\lambda,p,\sigma_t)}{I_i(\lambda,p)}\cdot\frac{R(\lambda,p)}{R(\lambda,\sigma_t)}\right] \\ &= \frac{1}{2}[\tau(\lambda,s,\sigma_t) + \tau(\lambda,p,\sigma_t)]. \end{aligned} \tag{A46}$$

As discussed above for Eq. (A42), Eq. (A46) is valid only if the receiver is able to discriminate the polarization of the radiant flux or there is no change in the polarization of the radiant flux that is transmitted by the sample. Applying the same procedure to Eqs. (A34), (A39), and (A40) yields

$$\begin{aligned}\tau(\lambda,u,t) &= \frac{1}{2}\left[\begin{array}{l}\dfrac{I_t(\lambda,s,s)}{I_i(\lambda,s)}\cdot\dfrac{R(\lambda,s)}{R(\lambda,s)} + \dfrac{I_t(\lambda,s,p)}{I_i(\lambda,s)}\cdot\dfrac{R(\lambda,s)}{R(\lambda,p)} + \\ \dfrac{I_t(\lambda,p,s)}{I_i(\lambda,p)}\cdot\dfrac{R(\lambda,p)}{R(\lambda,s)} + \dfrac{I_t(\lambda,p,p)}{I_i(\lambda,p)}\cdot\dfrac{R(\lambda,p)}{R(\lambda,p)}\end{array}\right] \\ &= \frac{1}{2}[\tau(\lambda,s,s) + \tau(\lambda,s,p) + \tau(\lambda,p,s) + \tau(\lambda,p,p)] \\ &= \frac{1}{2}[\tau(\lambda,s,t) + \tau(\lambda,p,t)]. \end{aligned} \tag{A47}$$

As with Eq. (A42), this expression is valid if the receiver is able to discriminate the polarization of the radiant flux. If there is no change in the polarization of the radiant flux that is transmitted by the sample, or the receiver is not sensitive to the polarization of the radiant flux, then Eq. (A47) reduces to

$$\tau(\lambda, u, t) = \frac{1}{2}\left[\frac{I_t(\lambda, s, t)}{I_i(\lambda, s)} + \frac{I_t(\lambda, p, t)}{I_i(\lambda, p)}\right] \quad (A48)$$
$$= \frac{1}{2}\left[\tau(\lambda, s, t) + \tau(\lambda, p, t)\right].$$

If the receiver is sensitive to polarization, but the ratio of the responsivities to the two orthogonal linear polarizations s and p is not known, $\tau(\lambda, \sigma_i, t)$ and $\tau(\lambda, u, t)$ can still be determined, but with an increase in the number of required measurements. A set of measurements is obtained with the polarization-sensitive component of the receiver rotated 0° from its normal position, then a second set of measurements is performed at a rotation of 90° about the axis of the radiant flux. This introduces an additional variable $r = 0°$ or 90° for the responsivity and hence the current. Summing the measured currents for the case of polarized incident radiant flux and total transmitted radiant flux yields

$$\frac{I_t(\lambda, \sigma_i, t, 0°) + I_t(\lambda, \sigma_i, t, 90°)}{I_i(\lambda, \sigma_i, 0°) + I_i(\lambda, \sigma_i, 90°)}$$
$$= \frac{I_t(\lambda, \sigma_i, s, 0°) + I_t(\lambda, \sigma_i, p, 0°) + I_t(\lambda, \sigma_i, s, 90°) + I_t(\lambda, \sigma_i, p, 90°)}{I_i(\lambda, \sigma_i, 0°) + I_i(\lambda, \sigma_i, 90°)} \quad (A49)$$
$$= \frac{\left[R(\lambda, s, 0°) + R(\lambda, s, 90°)\right] \cdot \Phi_t(\lambda, \sigma_i, s) + \left[R(\lambda, p, 0°) + R(\lambda, p, 90°)\right] \cdot \Phi_t(\lambda, \sigma_i, p)}{\left[R(\lambda, \sigma_i, 0°) + R(\lambda, \sigma_i, 90°)\right] \cdot \Phi_i(\lambda, \sigma_i)}$$

Now, while $R(\lambda, \sigma_t, 0°) \neq R(\lambda, \sigma_t, 90°)$ because of the sensitivity of the receiver to polarization, $R(\lambda, s, 0°) = R(\lambda, p, 90°) = R_1$ and $R(\lambda, p, 0°) = R(\lambda, s, 90°) = R_2$ because of the orthogonality of the linear polarizations and the 90° rotation of the receiver. Therefore, Eq. (A49) reduces to

$$\frac{I_t(\lambda, \sigma_i, t, 0°) + I_t(\lambda, \sigma_i, t, 90°)}{I_i(\lambda, \sigma_i, 0°) + I_i(\lambda, \sigma_i, 90°)} = \frac{\left[R_1 + R_2\right] \cdot \Phi_t(\lambda, \sigma_i, s) + \left[R_1 + R_2\right] \cdot \Phi_t(\lambda, \sigma_i, p)}{\left[R_1 + R_2\right] \cdot \Phi_i(\lambda, \sigma_i)}$$
$$= \frac{\Phi_t(\lambda, \sigma_i, t)}{\Phi_i(\lambda, \sigma_i)} \quad (A50)$$
$$= \tau(\lambda, \sigma_i, t).$$

Similarly,

$$\frac{I_t(\lambda, s, t, 0°) + I_t(\lambda, p, t, 0°) + I_t(\lambda, s, t, 90°) + I_t(\lambda, p, t, 90°)}{I_i(\lambda, s, 0°) + I_i(\lambda, p, 0°) + I_i(\lambda, s, 90°) + I_i(\lambda, p, 90°)} = \frac{1}{2}\left[\tau(\lambda, s, t) + \tau(\lambda, p, t)\right] \quad (A51)$$
$$= \tau(\lambda, u, t).$$

The current from the optical detector is usually not measured, but is converted to a voltage S [V] for measurement, given by

$$S(\lambda,\sigma) = I(\lambda,\sigma) \cdot G ,\qquad(A52)$$

where G [V/A] is the gain of the transimpedance amplifier. For the RTS, the receiver is insensitive to the polarization of the radiant flux, so the total transmitted flux is measured. Sets of signals are recorded as a function of wavelength and polarization of the incident radiant flux twice for the clear position and once for the sample position of the sample carriage. In addition, dark signals are recorded for each position. A single recorded signal is given by $S(\lambda, \sigma, i)$, where λ and σ are the wavelength and polarization of the incident radiant flux, respectively, and i designates the number in the set. The average signal is given by

$$S(\lambda,\sigma) = \frac{1}{n}\sum_{i=1}^{n} S(\lambda,\sigma,i) .\qquad(A53)$$

With the sample carriage at the clear position, the first average signal is designated by S_{i1}, the second by S_{i2}, and the dark signal by S_{id}. At the sample position, the signal is designated by S_{t1}, and the dark signal by S_{td}. The net signals are obtained by subtracting the dark signals, so

$$S_i(\lambda,\sigma) = \frac{1}{2}\left[S_{i1}(\lambda,\sigma) + S_{i2}(\lambda,\sigma) - 2S_{id}(\lambda,\sigma)\right] \text{ and}\qquad(A54)$$

$$S_t(\lambda,\sigma) = S_{t1}(\lambda,\sigma) - S_{td}(\lambda,\sigma) .\qquad(A55)$$

Using the expressions for the signals, Eqs. (A52) to (A55) with Eq. (A44) yields the measurements equation for the total transmitted radiant flux for a given wavelength and polarization of the incident radiant flux, given by

$$\tau(\lambda,\sigma_i,t) = \frac{S_t(\lambda,\sigma_i)}{S_i(\lambda,\sigma_i)} \cdot \frac{G_i(\lambda,\sigma_i)}{G_t(\lambda,\sigma_i)} ,\qquad(A56)$$

where G_t and G_i are the gains of the transimpedance amplifier when measuring the transmitted and incident radiant fluxes, respectively. The measurement equation for the total transmitted radiant flux at a given wavelength for unpolarized incident radiant flux is derived from Eqs. (A48) and (A56) and is given by

$$\tau(\lambda,u,t) = \frac{1}{2}\left[\frac{S_t(\lambda,s)}{S_i(\lambda,s)} \cdot \frac{G_i(\lambda,s)}{G_t(\lambda,s)} + \frac{S_t(\lambda,p)}{S_i(\lambda,p)} \cdot \frac{G_i(\lambda,p)}{G_t(\lambda,p)}\right].\qquad(A57)$$

In the calibration reports, this quantity is rewritten in a simpler form as

$$\tau(\lambda) = \frac{\tau(\lambda,0°) + \tau(\lambda,90°)}{2} ,\qquad(A58)$$

where

$$\tau(\lambda, \sigma_i) = \frac{S_t(\lambda, \sigma_i)}{S_i(\lambda, \sigma_i)} \cdot \frac{G_i(\lambda, \sigma_i)}{G_t(\lambda, \sigma_i)} = \frac{S_t(\lambda, \sigma_i)}{S_i(\lambda, \sigma_i)} \; , \tag{A59}$$

because for most calibrations, the gains, $G_t(\lambda, \sigma) \sim G_i(\lambda, \sigma)$, do not vary much.

Appendix B: Sample Report

REPORT OF CALIBRATION
38061S Regular Spectral Transmittance

for

Six Neutral Density Filters

Submitted by:

Any Company, Inc.
Attn.: Ms. Jane Doe
123 Calibration Street
Measurement City, MD 20800-1234

(See your Purchase Order No. 12345, dated January 1, 2010)

1. Description of Calibration Items

Six neutral density filters, 51 mm square, serial numbers F1-001, F1-002, F1-003, F2-001, F2-002, and F2-003, submitted by Any Company, Inc.

2. Description of Calibration

The regular spectral transmittance τ is the ratio of the radiant flux transmitted by the item, without diffusion, to the radiant flux incident on the item at normal incidence as a function of wavelength. For normal incidence, the axis of illumination is parallel to the normal of the front surface of the item.

The items were measured using the NIST Reference Transmittance Spectrophotometer. This instrument has undergone significant modifications for automation, but the design is similar to the one described in [1]. A spherical mirror focuses radiant flux from a source onto the entrance slit of a prism-grating monochromator. The beam emerging from the exit slit of the monochromator is collimated by an off-axis parabolic mirror, passes through a Glan-Taylor polarizer, and is incident on an iris to provide a circular illumination beam. The beam passes through the sample carriage, is collected by a fold mirror, and is focused by a spherical mirror into an averaging sphere. A signal proportional to the radiant flux of the beam is measured by an optical detector attached to the averaging sphere. The sample carriage consists of two incident positions for the beam: clear and sample. In each position a shutter is used as a light block for a dark signal. At each wavelength and polarization, signals are measured from the clear, dark, sample, dark, and clear positions, in order. Signals from the clear and sample positions are proportional to the incident and transmitted radiant fluxes, respectively. Net signals for the clear and sample positions are obtained by subtracting the dark position signals.

Individual items were cleaned with an air bulb and mounted in a commercial lens holder with the illumination beam centered on the front surface of the item at normal incidence. This was achieved by adjusting tilts and translations of the sample holder and carriage until a laser beam at a wavelength of 632.8 nm, propagating collinear to the beam from the monochromator, was centered on the front of the item and retroreflected.

The sampling aperture, defined by the illumination beam, had a diameter of 20 mm and was located at the center of the calibration item. The maximum deviation of any ray within the illumination beam from the normal direction was 0.81°. The maximum deviation of any ray within the receiver beam from the normal direction was 2.7°.

All items were measured at wavelengths from 380 nm to 770 nm every 10 nm. The spectral bandwidth of the illumination beam was 3 nm. The source was a QTH incandescent lamp, and the detector was a silicon photodiode. Each item was measured three times at two polarizer orientations rotated 90° from each other about the direction of the illumination beam.

3. Results of Calibration

The transmittance τ at each wavelength λ and polarization σ is given by

$$\tau(\lambda,\sigma) = \frac{S_t(\lambda,\sigma)}{S_i(\lambda,\sigma)} ,$$

where S_t is the net signal from the transmitted radiant flux in the sample position and S_i is the net signal from the incident radiant flux in the clear position. The regular spectral transmittance for an unpolarized illumination beam is given by

$$\tau(\lambda) = \frac{\tau(\lambda,0°) + \tau(\lambda,90°)}{2} .$$

The final regular spectral transmittance was obtained by averaging the values from both polarizations and is plotted in Fig. B1. The certified regular spectral transmittances of the calibration items, measured according to the details given above, are given in Tables B1 to B6.

Uncertainties were calculated according to the procedures outlined in [2]. Sources of uncertainty due to random effects are source stability and detector noise. The uncertainty contribution caused by these sources was evaluated from the standard deviation of repeat measurements of each item.

Sources of uncertainty due to systematic effects are the wavelength of the monochromator and linearity of the receiver. The uncertainty contribution caused by the wavelength uncertainty was evaluated from the derivative of the transmittance. The uncertainty contribution caused by linearity, which was determined in the characterization of the RTS facility, includes effects from both the detector and the signal electronics. All uncertainty components were assumed to have normal probability distributions.

The sources of uncertainty and uncertainty contributions are given in Tables B7 to B12 for the items. The expanded uncertainty in regular spectral transmittance is obtained from the root-sum-square of the uncertainty contributions multiplied by a coverage factor $k = 2$.

4. General Information

1) The items were measured in the "as received" condition, after cleaning with an air bulb.

2) The data in Tables B1 to B6 apply only to the central 20 mm diameter area of the calibration item.

3) This calibration report may not be reproduced except in full without the written consent of this Laboratory.

Prepared by: Approved by:

Catherine C. Cooksey   Eric L. Shirley
Optical Technology Division   For the Director,
Physical Measurement Laboratory   National Institute of
(301) 975-6208   Standards and Technology
                                                    (301) 975-2349

References:

[1] K. L. Eckerle, J. J. Hsia, K. D. Mielenz, and V. R. Weidner, "NBS Measurement Services: Regular Spectral Transmittance," NBS Special Publication 250-6 (1987).

[2] B. N. Taylor and C. E. Kuyatt, "Guidelines for Evaluating and Expressing the Uncertainty of NIST Measurement Results," NIST Technical Note 1297 (1994).

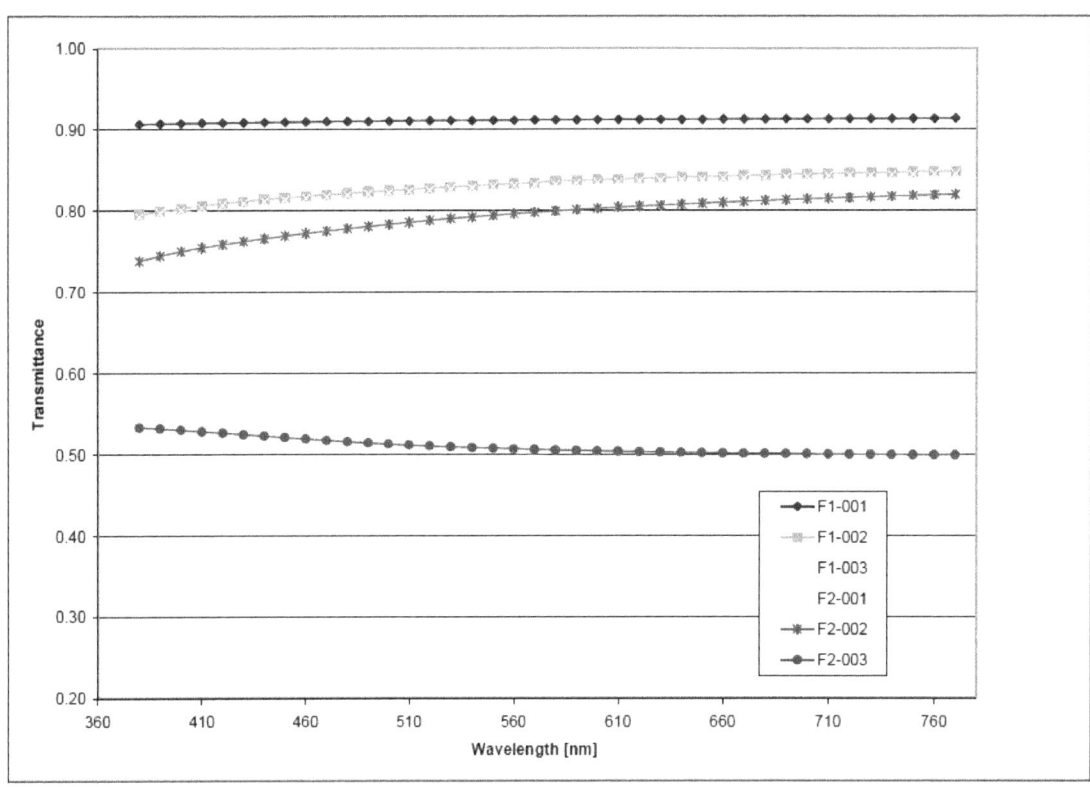

Figure B1. Regular spectral transmittance τ as a function of wavelength λ of neutral density filters.

Table B1. Regular spectral transmittance τ as a function of wavelength λ of a neutral density filter, serial number F1-001.

λ [nm]	τ	λ [nm]	τ
380	0.9067	580	0.9123
390	0.9077	590	0.9124
400	0.9082	600	0.9125
410	0.9087	610	0.9126
420	0.9090	620	0.9126
430	0.9092	630	0.9126
440	0.9095	640	0.9127
450	0.9097	650	0.9128
460	0.9101	660	0.9130
470	0.9104	670	0.9131
480	0.9106	680	0.9132
490	0.9107	690	0.9133
500	0.9110	700	0.9134
510	0.9112	710	0.9134
520	0.9114	720	0.9136
530	0.9116	730	0.9136
540	0.9117	740	0.9137
550	0.9120	750	0.9138
560	0.9120	760	0.9138
570	0.9121	770	0.9140

Table B2. Regular spectral transmittance τ as a function of wavelength λ of a neutral density filter, serial number F1-002.

λ [nm]	τ	λ [nm]	τ
380	0.7956	580	0.8367
390	0.7999	590	0.8376
400	0.8034	600	0.8386
410	0.8066	610	0.8394
420	0.8095	620	0.8402
430	0.8119	630	0.8410
440	0.8143	640	0.8418
450	0.8164	650	0.8425
460	0.8183	660	0.8432
470	0.8201	670	0.8438
480	0.8218	680	0.8445
490	0.8235	690	0.8451
500	0.8251	700	0.8457
510	0.8267	710	0.8463
520	0.8281	720	0.8468
530	0.8295	730	0.8474
540	0.8310	740	0.8478
550	0.8322	750	0.8484
560	0.8334	760	0.8489
570	0.8346	770	0.8491

Table B3. Regular spectral transmittance τ as a function of wavelength λ of a neutral density filter, serial number F1-003.

λ [nm]	τ	λ [nm]	τ
380	0.6582	580	0.6694
390	0.6600	590	0.6697
400	0.6612	600	0.6700
410	0.6623	610	0.6703
420	0.6631	620	0.6707
430	0.6637	630	0.6709
440	0.6642	640	0.6712
450	0.6646	650	0.6716
460	0.6650	660	0.6720
470	0.6653	670	0.6723
480	0.6656	680	0.6726
490	0.6659	690	0.6729
500	0.6664	700	0.6732
510	0.6668	710	0.6735
520	0.6671	720	0.6738
530	0.6675	730	0.6742
540	0.6680	740	0.6745
550	0.6683	750	0.6747
560	0.6687	760	0.6750
570	0.6690	770	0.6753

Table B4. Regular spectral transmittance τ as a function of wavelength λ of a neutral density filter, serial number F2-001.

λ [nm]	τ	λ [nm]	τ
380	0.7931	580	0.8330
390	0.7973	590	0.8341
400	0.8008	600	0.8351
410	0.8041	610	0.8359
420	0.8069	620	0.8368
430	0.8094	630	0.8377
440	0.8116	640	0.8383
450	0.8137	650	0.8391
460	0.8157	660	0.8399
470	0.8175	670	0.8405
480	0.8192	680	0.8412
490	0.8209	690	0.8418
500	0.8225	700	0.8425
510	0.8240	710	0.8431
520	0.8254	720	0.8435
530	0.8269	730	0.8441
540	0.8282	740	0.8446
550	0.8296	750	0.8451
560	0.8308	760	0.8457
570	0.8320	770	0.8462

Table B5. Regular spectral transmittance τ as a function of wavelength λ of a neutral density filter, serial number F2-002.

λ [nm]	τ	λ [nm]	τ
380	0.7384	580	0.7999
390	0.7451	590	0.8015
400	0.7504	600	0.8029
410	0.7549	610	0.8044
420	0.7589	620	0.8057
430	0.7626	630	0.8069
440	0.7661	640	0.8081
450	0.7694	650	0.8093
460	0.7727	660	0.8105
470	0.7756	670	0.8115
480	0.7785	680	0.8126
490	0.7811	690	0.8136
500	0.7837	700	0.8146
510	0.7860	710	0.8155
520	0.7885	720	0.8164
530	0.7906	730	0.8173
540	0.7927	740	0.8182
550	0.7946	750	0.8191
560	0.7964	760	0.8199
570	0.7983	770	0.8207

Table B6. Regular spectral transmittance τ as a function of wavelength λ of a neutral density filter, serial number F2-003.

λ [nm]	τ	λ [nm]	τ
380	0.5337	580	0.5059
390	0.5323	590	0.5053
400	0.5307	600	0.5047
410	0.5289	610	0.5042
420	0.5270	620	0.5037
430	0.5250	630	0.5032
440	0.5232	640	0.5028
450	0.5213	650	0.5024
460	0.5195	660	0.5020
470	0.5178	670	0.5018
480	0.5162	680	0.5015
490	0.5148	690	0.5012
500	0.5134	700	0.5010
510	0.5122	710	0.5008
520	0.5110	720	0.5005
530	0.5100	730	0.5003
540	0.5091	740	0.5001
550	0.5082	750	0.4999
560	0.5074	760	0.4998
570	0.5066	770	0.4996

Table B7. Uncertainty contributions and expanded uncertainty ($k = 2$) of the regular transmittance of a neutral density glass filter, serial number F1-001.

Source of Uncertainty	Standard Uncertainty	Uncertainty Contribution
Wavelength	0.1 nm	0.0000
Receiver Linearity	0.05 %	0.0005
Repeatability	0.0001	0.0001
		Expanded Uncertainty ($k = 2$)
		0.0010

Table B8. Uncertainty contributions and expanded uncertainty ($k = 2$) of the regular transmittance of a neutral density glass filter, serial number F1-002.

Source of Uncertainty	Standard Uncertainty	Uncertainty Contribution
Wavelength	0.1 nm	0.0000
Receiver Linearity	0.05 %	0.0004
Repeatability	0.0001	0.0002
		Expanded Uncertainty ($k = 2$)
		0.0009

Table B9. Uncertainty contributions and expanded uncertainty ($k = 2$) of the regular transmittance of a neutral density glass filter, serial number F1-003.

Source of Uncertainty	Standard Uncertainty	Uncertainty Contribution
Wavelength	0.1 nm	0.0000
Receiver Linearity	0.05 %	0.0003
Repeatability	0.0001	0.0001
		Expanded Uncertainty ($k = 2$)
		0.0008

Table B10. Uncertainty contributions and expanded uncertainty ($k = 2$) of the regular transmittance of a neutral density glass filter, serial number F2-001.

Source of Uncertainty	Standard Uncertainty	Uncertainty Contribution
Wavelength	0.1 nm	0.0000
Receiver Linearity	0.05 %	0.0004
Repeatability	0.0001	0.0001
		Expanded Uncertainty ($k = 2$)
		0.0008

Table B11. Uncertainty contributions and expanded uncertainty ($k = 2$) of the regular transmittance of a neutral density glass filter, serial number F2-002.

Source of Uncertainty	Standard Uncertainty	Uncertainty Contribution
Wavelength	0.1 nm	0.0000
Receiver Linearity	0.05 %	0.0004
Repeatability	0.0001	0.0001
		Expanded Uncertainty ($k = 2$)
		0.0008

Table B12. Uncertainty contributions and expanded uncertainty ($k = 2$) of the regular transmittance of a neutral density glass filter, serial number F2-003.

Source of Uncertainty	Standard Uncertainty	Uncertainty Contribution
Wavelength	0.1 nm	0.0000
Receiver Linearity	0.05 %	0.0003
Repeatability	0.0001	0.0001
		Expanded Uncertainty ($k = 2$)
		0.0006

Appendix C: Sample Data

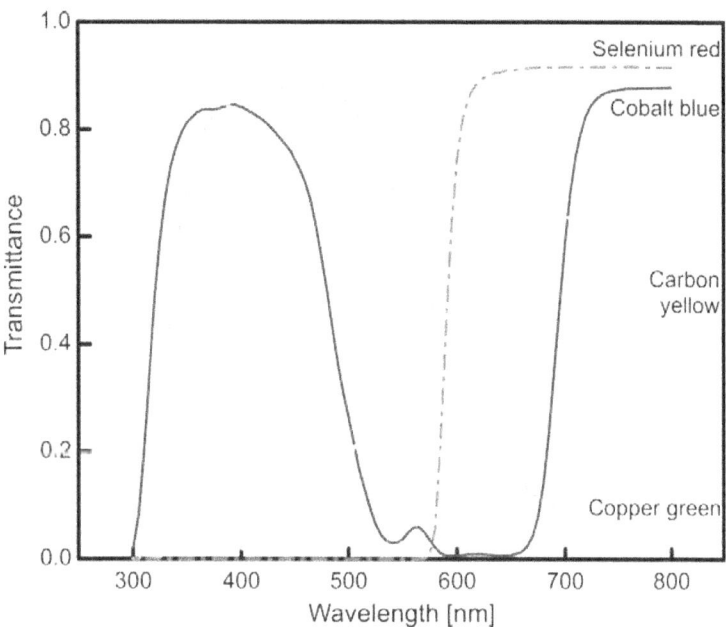

Figure C1. Transmittance as a function of wavelength of the color filters.

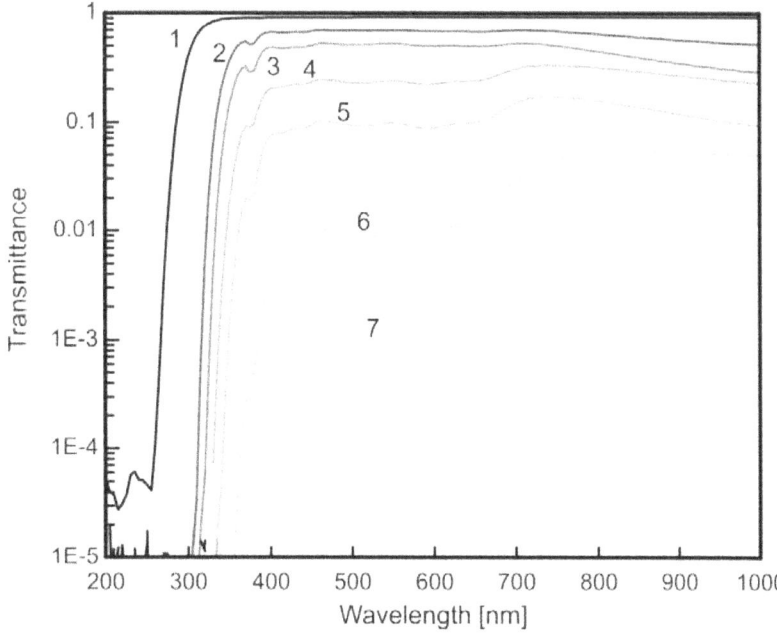

Figure C2. Transmittance, on a logarithmic scale, as a function of wavelength of the MAP (Measurement Assurance Program) filters [15]. Nominal transmittance values and materials for the MAP filters are given below in Table C3.

Table C3. Nominal transmittance values at 548.5 nm and materials for the MAP filters.

Filter No.	Nominal τ	Material
1-1	0.9	Borosilicate crown glass
1-2	0.7	Schott NG-11
1-3	0.5	Schott NG-11
1-4	0.25	Schott NG-4
1-5	0.10	Schott NG-4
1-6	0.01	Schott NG-9
1-7	0.001	Schott NG-9

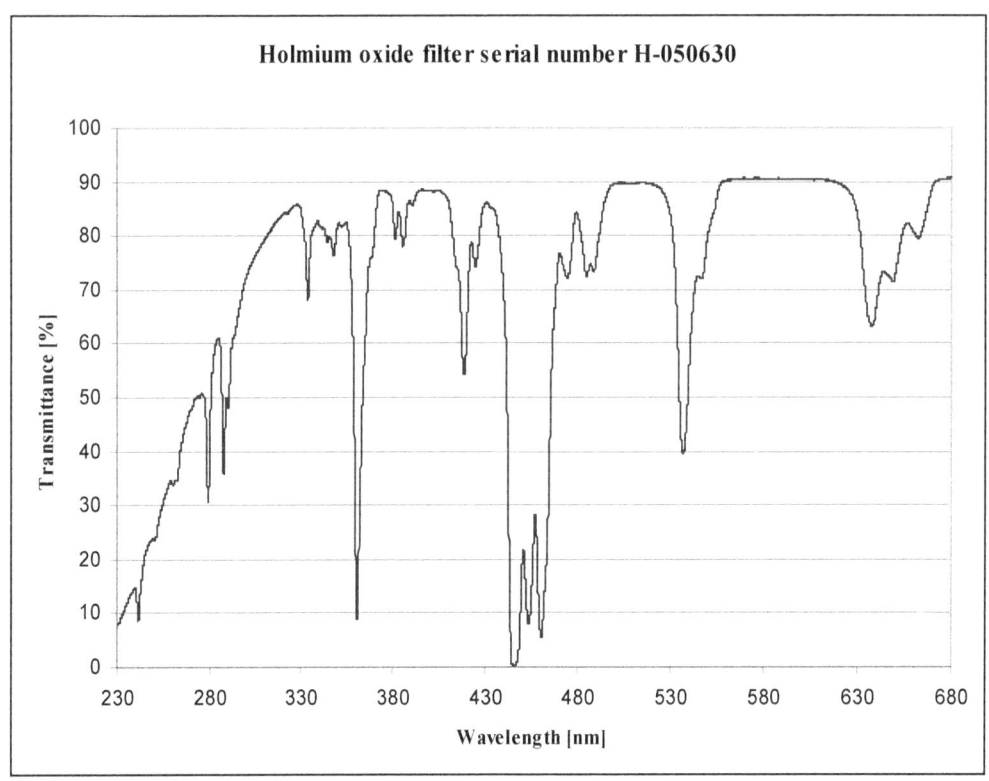

Figure C4. Measured spectral transmittance of the holmium oxide glass filter for a spectral bandwidth of 1 nm.

Appendix D: How to Request Regular Spectral Transmittance Calibrations

1. Prepare a purchase order with the following (discuss with technical contact prior to submitting a formal request):
    a. Service ID number requested (include range and points): 38061S.
    b. Clear identification of calibration item(s), including any model or serial numbers.
    c. Company name and address and name and phone number of contact person to receive the calibration report(s).
    d. Billing and shipping addresses.
    e. Return shipping instructions (prepay and add, COD, charge to account with shipper)
        i. Test fee does not include shipping; customers are responsible for all shipping costs.
        ii. Without instructions, NIST will return test item by common carrier, collect, and uninsured.
    f. Special handling instructions.

2. Send the purchase order to:
    Calibration Services
    National Institute of Standards and Technology
    100 Bureau Drive, Stop 2330
    Gaithersburg, MD 20899-2330
    Tel.: (301) 975-2002
    FAX: (301) 869-3548
    E-mail: calibrations@nist.gov

3. Send the test item(s) to:
    Catherine C. Cooksey
    National Institute of Standards and Technology
    100 Bureau Drive, Stop 8442
    Gaithersburg, MD 20899-8442
    Tel.: (301) 975-6208
    FAX: (301) 975-6991
    E-mail: catherine.cooksey@nist.gov
    WWW: http://www.nist.gov/pml/div685/ grp03/staff/cooksey.cfm

After receipt of purchase order, calibration services are scheduled to be completed in 90 days. NIST policy requires prepayment of all calibration services performed for non-U.S. organizations. Please contact the Calibration Program office to arrange payment. Despite best efforts sample may in rare instances incur damage during routine handling. Customers should consider a contingency or replacement plan to minimize any disruption prior to submitting samples.

Please note that the Analytical Chemistry Division calibrates the following SRMs: 930, 1930, 2930, and 2031. For recalibration of these SRMs, please contact:

Melody Smith
Phone: 301-975-4115
FAX: 301-977-0587
Email: melody.smith@nist.gov